MODERN PETROCHEMICALS

Ali El-Shekeil

Professor of Chemistry

Sana'a University, Yemen

2013

Table of Contents

TABLE OF CONTENTS I

TABLE OF FIGURES IV

TABLE OF TABLES...V

INTRODUCTION VII

CHAPTER ONE   OIL AS A SOURCE OF ENERGY AND RAW MATERIALS 1

CHAPTER TWO   PETROCHEMICALS FROM SYNTHESIS GAS 9

SYNTHESIS GAS.. 10

AMMONIA 14

UREA 17

METHYLAMINES 17

NITRIC ACID.. 18

MELAMINE.. 19

METHANOL 20

FORMALDEHYDE 22

FORMIC ACID 23

GAS TO LIQUIDS 24

HYDROFORMYLATION OF OLEFINS 27

OXO PRODUCTS... 30

OLEFIN CARBONYLATION 32

CHAPTER THREE   PETROCHEMICALS FROM ETHYLENE 35

ETHYLENE 36

MECHANISM OF THERMAL CRACKING.. 43

POLYETHYLENE.. 47

METALLOCENES 51

ETHYL BENZENE... 53

STYRENE 55

POLYSTYRENE 58

LINEAR ALKYL BENZENE 59

LINEAR ALPHA OLEFINS 61

SABLIN NEW LINEAR ALPHA OLEFINS 61

ETHANOL (ETHYL ALCOHOL)... 64

ACETALDEHYDE 66

ACETIC ACID 69

Synthesis of acetic acid by oxidation of acetaldehyde 69

Synthesis of acetic acid by oxidation of alkanes and alkenes 70

Synthesis of acetic acid from methanol 71

i

Uses of acetic acid.. 73

ETHYL ACETATE 74

ETHYLENE OXIDE.. 74

ETHYLENE GLYCOL.. 78

ETHYLENE CHLORIDE.. 79

ETHANOLAMINES 81

ETHYLENE BROMIDE 82

VINYL CHLORIDE 82

1,1,2-TRICHLOROETHYLENE 86

PERCHLOROETHYLENE 88

ETHYL CHLORIDE 89

VINYL FLUORIDE 90

TETRAFLOUROETHYLENE 91

VINYL ACETATE... 91

CHAPTER FOUR   PETROCHEMICALS FROM PROPYLENE.............................95

PROPYLENE 96

POLYPROPYLENE... 100

ISOPROPYL ALCOHOL 101

ACETONE 102

METHYL ISOBUTYL KETONE 105

ACRYLIC ACID AND ITS ESTERS... 106

METHACRYLIC ACID AND ITS ESTERS.. 108

ACROLEIN 112

PROPYLENE OXIDE.. 113

ALLYL CHLORIDE 115

ALLYL ALCOHOL.. 116

GLYCERIN.. 118

ACRYLONITRILE 119

DI-, TRI- AND TETRAPROPYLENE.. 120

CHAPTER FIVE   PETROCHEMICALS FROM BUTENES AND BUTADIENE 123

BUTENES 124

HIGHER OLEFINS... 125

BRANCHED HIGHER OLEFINS... 127

BUTADIENE 129

ISOPRENE.. 133

BUTAN-1,4-DIOL... 134

1-BUTENE.. 137

N- AND I-BUTYRALDEHYDE.. 138

BUTYL ALCOHOLS... 139

VINYL ETHERS 141

CHAPTER SIX   PETROCHEMICALS FROM AROMATICS 143

PETROCHEMICALS FROM AROMATICS 144
    Catalytic reforming 145
    Continuous catalytic reforming 147
    Cyclar process 148
    Pyrolysis gasoline 149
    SEPARATION PROCESSES 150
    Solvent extraction 151
    Azeotropic distillation 152
    Fractional distillation 152
    Extractive distillation 152
    Crystallization 154
    Solvent-solvent extraction 154
PETROCHEMICALS FROM BENZENE 155
    Bisphenol A 159
    Cumene 161
    Linear alkyl benzene 163
    Cyclohexane 164
    Phenol 165
    Nitrobenzene 169
    Maleic anhydride 170
PETROCHEMICALS FROM TOLUENE 173
    Benzoic Acid 174
    Trinitrotoluene 175
    ε-caprolactam 175
PETROCHEMICALS FROM XYLENES 178
    Isomerization of Xylenes 178
    Phthalic Anhydride 179
    Terephthalic Acid 181
    Dimethyl Terephthalate 187

REFERENCES (GENERAL) 192

Table of Figures

Figure 1-1: The main sources of primary petrochemicals are natural gas and crude oil.. 2

Figure 1-2: The primary petrochemicals are produced from naphtha through steam cracking for ethylene, propylene and butenes or by catalytic reforming for the preparation of the three aromatics (BTX)................................. 3

Figure 1-3: The top proven reserves of oil as of January 1, 2011. (Billion tons). Source: Oil & Gas Journal. 5

Figure 1-4: The top global proven reserves of natural gas as of January 1, 2011. (Trillion cubic feet) 6

Figure 2-1: The main petrochemical derivatives of synthesis gas. 10

Figure 2-2: The main petrochemical derivatives of methanol. 22

Figure 2-3: The gas-to-liquids process.. 25

Figure 3-1: The most prominent petrochemical derivatives of ethylene. 37

Figure 3-2: Additions of ethylene complexes in the World by region between 1998 and 2003 were 28 million tons. 42

Figure 3-3: The volume of the global demand for polymers, 2005. 49

Figure 3-4: Metallocenes are the second generation of Ziegler-Natta catalysts. 52

Figure 3-5: SABLIN new synthesis of alpha olefins.. 62

Figure 3-6: The most notable byproducts of acetaldehyde................................. 68

Figure 4-1: The most important petrochemical derivatives of propylene.............. 99

Figure 4-2: The main uses of acrolein.. 112

Figure 5-1: The most important industrial uses of isobutene.............................. 124

Figure 5-2: The main petrochemical derivatives of butadiene........................... 133

Figure 6-1: The syntheses of aromatics from oil derivatives............................. 144

Figure 6-2: The global benzene capacity by process, 2010. 155

Figure 6-3: The most important industrial derivatives of benzene. 156

Figure 6-4: The World consumption of benzene, 2010. 158

Figure 6-5: The main derivatives of phenol... 168

Figure 6-6: The utmost significant uses of aniline.. 170

Figure 6-7: The principal products derived from maleic anhydride. 172

Figure 6-8: The utmost essential industrial derivatives of toluene. 173

Table of Tables

Table 1-1: The main pathways of primary petrochemicals into petrochemical intermediates, derivatives and performance petrochemicals and their market end uses. 7

Table 1-2: The growth and share in the World production capacity of petrochemicals (million ton/year) in the Middle East between 2009 and 2015. 8

Table 2-1: Formation of carbon dioxide as a by-product in the preparation of synthesis gas. 13

Table 2-2: The most prominent industrial uses of synthesis gas. 13

Table 2-3: The most famous international ammonia technologies. 16

Table 2-4: The most notable characteristics and differences between the most famous companies and technologies for methanol production. 21

Table 2-5: The announced GTL projects (2002) and their capacities. 27

Table 2-6: The typical range of products in a GTL process. 27

Table 3-1: The most prominent global uses of ethylene (ratio %) in 2000. 37

Table 3-2: The production ceiling of ethylene using different feedstocks. 40

Table 3-3: Comparison between the four main global pioneer companies in the manufacture of ethylene complexes in terms of yields obtained using various feedstocks as weight % and energy consumption per kg of ethylene produced. 41

Table 3-4: The ideal distribution for the production of ethylene and propylene using different feed stocks and technologies, fluid catalytic crackers, Coker, and gasoline pyrolysis. 43

Table 3-5: The most powerful technologies for steam cracking of ethane. 46

Table 3-6: The most prominent natural properties of polyethylenes. 51

Table 3-7: A comparative study of the most significant new technologies used in the production of polyethylene. 53

Table 3-8: The most serious effects of the second-generation metallocenes on the properties of polyethylenes. 53

Table 3-9: The distribution of alpha olefin products (weight %). 63

Table 3-10: The conditions of reaction of olefins with sulfuric acid, for the synthesis of alcohols. 65

Table 3-11: The most valuable uses of acetic acid in 2000 (Wt. %). 73

Table 3-12: The fundamental differences between the three technologies of ethylene oxide. 75

Table 3-13: The most valuable uses of ethylene oxide in 2007 (ratio %). 78

Table 3-14: The global uses of polyvinyl acetate in 2000 (Wt. %). 93

Table 4-1: The most important uses of propylene in the world in 2000 (Wt. %). ... 96

Table 4-2: A summary of the comparative demand of North America, Western Europe and South East Asia for propylene in 2003 (wt. %). 97

Table 4-3: An ideal distribution of products from Superflex Process using different feeds. 98

Table 4-4: The most significant differences and characteristics of the different polypropylene technologies... 100

Table 5-1: The composition of C₄ fraction produced from the steam cracking of naphtha and the catalytic cracking of diesel (weight %)................... 125

Table 5-2: Raw materials and type of process for producing branched olefins and respective products. 129

Table 5-3: The content of butadiene (weight %) in the steam cracking of a number of feedstocks. 130

Table 5-4: The international trend towards the production of synthetic rubber in the World (Wt.%). 132

Table 5-5: The global butadiene consumption by application, 2011 (ratio %)..... 132

Table 5-6: The most important uses of butane-1,4-diol. 136

Table 6-1: The yields of the continuous catalytic reforming process (weight%). 148

Table 6-2: The yields in the Cyclar process according to the feed (weight%). 149

Table 6-3: The typical distribution of the components of pyrolysis gasoline, catalytic reforming and Cyclar Process.. 150

Table 6-4: Some natural properties of C₈ fraction aromatics............................... 150

Table 6-5: The main aromatics extraction operations. 151

Table 6-6: Some solvents used in the extraction of aromatics............................. 153

Table 6-7: Examples of extractive distillation. 154

Table 6-8: The production yields of benzene and xylene by Dettol process. 158

Table 6-9: The typical global distribution of toluene uses, 2010......................... 174

Table 6-10: The typical distribution of the yields of Isomar and Parex complex. 179

vi

Introduction

Petrochemicals are the branch of chemicals produced partially or totally from petroleum or natural gas. The World was familiar with many of those chemicals, known recently as petrochemicals, before the advent of petroleum and natural gas, such as ethyl alcohol that was previously produced from the fermentation of sugars, starches and cellulosic materials. Ethyl alcohol was dehydrated to ethylene then polymerized to polyethylene. I fact, some petroleum poor countries like Peru and Pakistan prepared ethyl alcohol from rotten fruits then dehydrated it into ethylene and consequently polymerized into polyethylene to avoid importing petroleum and petrochemicals. Similarly, glycerin was extracted as a byproduct in the soap industry and the aromatics were distilled from coal tar. When oil was discovered, it was converted into the same old products using easier methods, in cheaper prices and vast quantities.

Petroleum, compared to other starting materials, is characterized by cheaper price, availability in large quantities, ease of transport, diversity of products and uses besides production of many hydrocarbons that can be used as feedstocks for the petrochemical industries.

Innovation had a non-forgettable role to play in the development of petrochemicals as a science and industry. Many materials were synthesized or extracted for academic purposes, then turned into industrial products in enormous quantities, due to the fast development of organic chemistry. Styrene, for example, was discovered in 1830 and exploited industrially after more than a century from that date.

Many other efforts and circumstances have combined to make the development of petrochemical industries a success such as the market demand, availability of suitable technologies and thinking minds.

Isopropyl alcohol was the first material derived from petroleum in the United States of America, shortly after the First World War, as a starting material for the production of acetone. Acetone was used by then in the manufacture of explosives and as a solvent in the aircraft industry for painting the wings textiles.

In 1926, Union Carbide produced ethylene glycol from ethylene, which was a milestone in petrochemicals industries. Ethylene glycol was used at the beginning as a non-boiling liquid in car radiators and to prevent freezing of water in the winter.

Car industry played a significant role in the growth and development of petrochemicals industries, not only as a consumer of petrochemicals but for the production of gasoline from crude oil. This leads to the production of a number of byproducts that became basic feedstocks in the petrochemicals industries.

The growth and development of petrochemicals industries continued steadily and doubled in the period 1950-1980 including the solvents, fibers, plastics, rubber, resins, and detergents among others. The global energy consumption also tripled between 1965 and 2011.

The petrochemicals industries introduced excellent alternatives to some natural materials such as cotton, wool and natural rubber, in massive quantities with extremely low prices, like polyester, styrene-butadiene rubber and polyacrylates, respectively.

Currently, the petrochemical products reached all applications of the modern citizen to the extent that it seems impossible to imagine living without them. The petrochemical products dominated all aspects of civilized life including building and structure, transportation, textiles and clothing, food, agriculture, industry and the everyday use.

Petrochemicals are divided into three types, primary petrochemicals, which include ethylene, propylene, butadiene and butenes, benzene, toluene and xylenes, together with methanol. These are the main bricks in the petrochemicals industries from which hundreds of petrochemicals are made; called the intermediate petrochemicals such as styrene and acrylonitrile.

The intermediate petrochemicals can be reacted to give the performance or end use petrochemicals such as polystyrene and polyacrylonitrile.

This book is designed to study the primary petrochemicals and their chemistry. The book starts by an introduction, followed by a chapter about oil as a source of energy and raw materials, and a chapter covering the industrial chemistry of synthesis gas. The main four chapters of the book are devoted to the industrial chemistry of ethylene, propylene, butenes and butadiene and finally the aromatics.

I have been teaching this course in petrochemicals at the Faculty of Science, The University of Sana'a, Yemen, since 1980 before collecting my notes into this book, hoping that it fulfills the aim of propagation of knowledge to the whole world.

Chapter One
Oil as a Source of Energy and Raw Materials

The crude oil provided the World with a wonderful source of energy and raw materials for the chemical and petrochemical industries. Oil is one of the three fossil fuels: oil, natural gas and coal. The price of oil, its availability and reliability set its appropriate use in the World. The concern about the low price of oil increased due to its role in providing a main source of energy and the industrial raw materials to enhance the global economy in the case of depletion of oil.

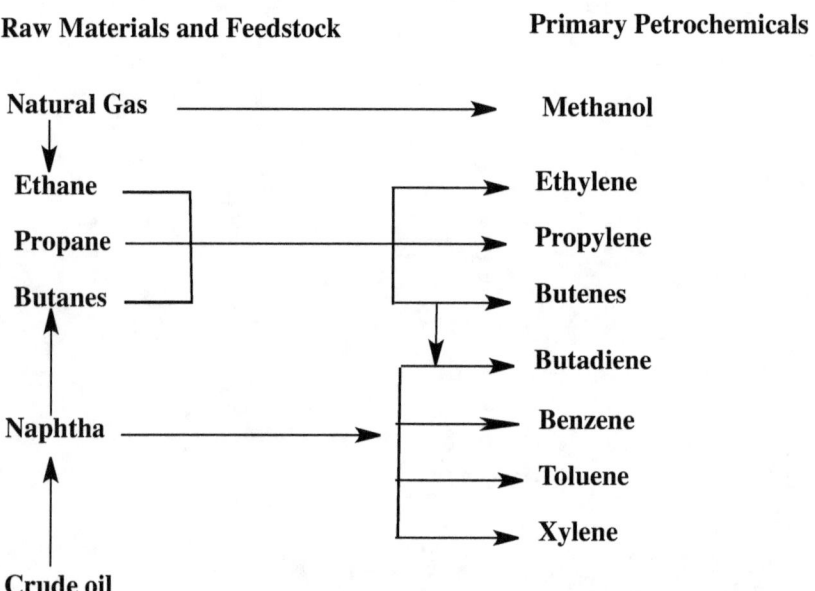

Figure 1-1: The main sources of primary petrochemicals are natural gas and crude oil.

In the opinion of all experts of the world, the importance of ensuring the availability of energy and raw materials, for the longest possible period of time is crucial. This is anticipated through extending the

2

life of the fossil fuel raw materials, replacement of raw materials dependent on fossil fuels and energy, and reliance on alternative and renewable energy sources, with considering the re-use of fossil fuel raw materials. This approach serves several purposes at the same time; to provide raw materials for chemical industries, energy from available and alternative sources, maintaining the cleanliness of the environment, reduction of heat of the atmosphere, mitigating the emission of waste gases, caused mainly by the fossil fuels, and finally, preserving the natural resources in the globe, and utilizing them in an optimal way.

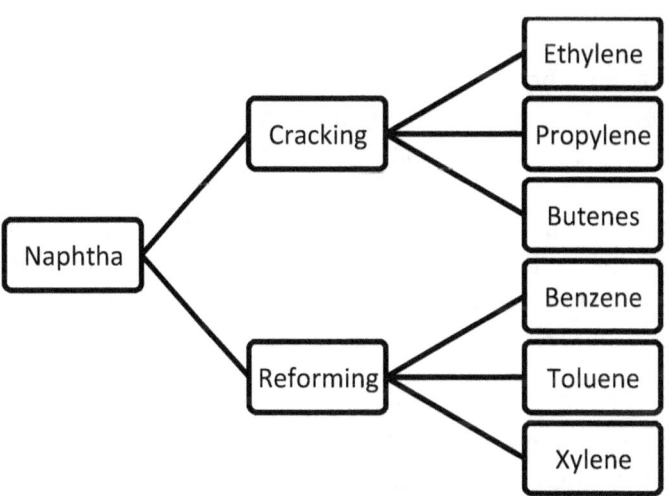

Figure 1-2: The primary petrochemicals are produced from naphtha through steam cracking for ethylene, propylene and butenes or by catalytic reforming for the preparation of the three aromatics (BTX).

Figure (1-1) illustrates the main sources of primary petrochemicals from natural gas and crude oil. The global energy consumption increased steadily with the increasing world population, and the rise in living standards of humanity. It approached 90 million barrels per day in 2012. The global energy consumption has tripled from 1965 to

2011. Figure (1-2) summarizes the primary petrochemicals production from naphtha through steam cracking for ethylene, propylene and butenes or by catalytic reforming for the preparation of the three aromatics (BTX).

The uses of primary energy can be divided essentially into three almost equal sections: one third for transportation, a third for industrial uses, and another third for domestic, agricultural and other uses. Oil and natural gas enjoy a set of irresistible properties in use as an energy source, such as economy, flexibility and excellent applications, in addition to ease of transportation and distribution. Thus, oil and natural gas will remain the utmost principal sources of energy in the near future.

The great demand for oil and natural gas called for further exploration, which has evolved with the continued development in various fields of technology, resulting in doubling of the global reserves of oil and natural gas. Although the majority of large oil fields had been previously discovered, exploration, however, covered the entire globe to reach the deep oceans, and revealed the hidden fields, including the medium and small ones. Scientists also discovered other sources of crude oil from oil shale, tar sands and oil sands.

The scientists expect that the world's reserves of oil and natural gas, in addition to synthetic oil extracted from oil sands, oil shale and tar sands, to cover the needs of the World's energy consumption until the end of the current century, bearing in mind the possibility of extracting the entire inventory, besides some innovative developments in the area of renewable and alternative energies. This would leave some fossil resources for use as raw materials in the oil and petrochemical industry. Figures 1-3 and 1-4 show the top proven reserves of oil and gas by 2011, respectively.

Natural gas and oil will become the primary sources of energy, which would lead to competition between their use in the energy area, or as valuable raw materials for chemical and petrochemical industries.

The main pathways of primary petrochemicals into petrochemical intermediates, derivatives and performance petrochemicals and their market end uses are shown in Table (1-1).

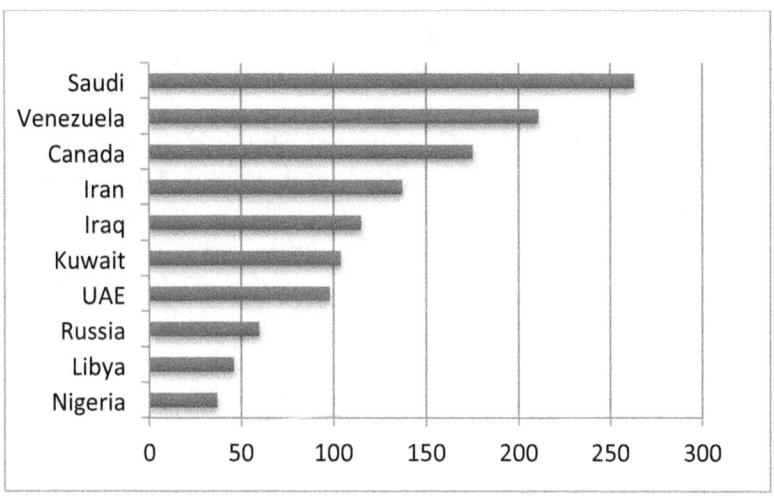

Figure 1-3: The top proven reserves of oil as of January 1, 2011. (Billion tons). Source: Oil & Gas Journal.

The renewable and alternative energy sources will be introduced steadily, to replace fossil fuels, particularly oil and natural gas over the medium term together with nuclear energy, especially in rich and developed countries. On the long run, solar energy and nuclear fusion would play a larger role in the energy sources. Nuclear fusion will solve all energy problems of the world for millions of years to come if the scientists have managed to understand and control it. That would leave the fossil fuels for the petrochemical industries and welfare of the humanity.

On the long run, other renewable energies will have a promising role, such as solar, wind and geothermal energies and hydropower. Until those techniques get developed to satisfy large projects, the World will depend on alternatives to oil and natural gas such as coal, nuclear energy and hydrogen energy, in particular.

It can be concluded from the past discussion that on the long run the field of energy will depend on renewable and alternative energy resources while fossil fuel resources of oil, natural gas and coal will provide raw materials for petrochemical industries.

Fundamental changes will occur in the sources of essential raw materials for the chemical industry and would take maximum advantage of the cheap sources of coal by gasification, and conversion to preliminary chemicals through hydrogenation and carbonization at low temperatures. Moreover, the World will return to the production of acetylene from calcium carbide, the technique that flourished directly before the advent of oil, but disappeared due to the lack of ability to compete economically. When the evolution of these operations is technically visible, coal will replace oil and natural gas as a source of raw materials. The future will verify this hypothesis.

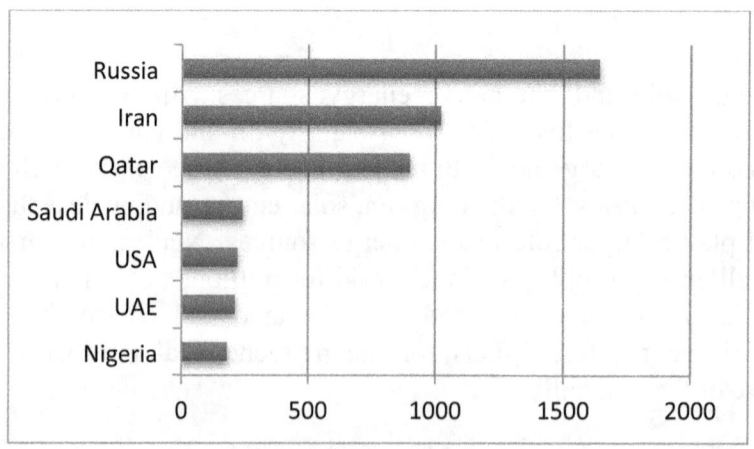

Figure 1-4: The top global proven reserves of natural gas as of January 1, 2011. (Trillion cubic feet)

Table 1-1: The main pathways of primary petrochemicals into petrochemical intermediates, derivatives and performance petrochemicals and their market end uses.

Primary Petrochemicals	Petrochemical Intermediates And Derivatives		Performance Petrochemicals	Major End Use
Methanol	Formaldehyde		Phenol formaldehyde resins	Plastics Adhesives
	Acetic Acid	Acetic anhydride	Cellulose acetate	Fibers
		Vinyl acetate	Polyvinyl acetate	Paper and textile sizing
Ethylene	Ethanol			Solvent, cosmetics + pharmaceuticals
	Ethylene oxide	Ethylene glycol		Coolant and fibers
	Ethylene dichloride	Vinyl chloride	Polyvinyl chloride	
	Ethylbenzene	Styrene	Polystyrene	Plastic Products
	Polyethylene resins			
Propylene	Cumene	Phenol and acetone		
			Polypropylene	Plastic products + fibers
	Isopropanol			Paint solvent
	Propylene oxide		Polyether/Polyols	Polyurethane foam products
Butadiene			Synthetic rubber + latex	Tires + rubber products
Benzene	Cyclohexane	Adipic acid	Nylon 66	Fibers

The Arabian Gulf is one of the greatest reserves of oil and gas in the World. A number of petrochemical factories have been installed. In 2009, it produced about 11.5 % of the petrochemicals in the world (Table (1-2)). It is expected to become one of the centers of gravity of this promising industry and produce more than 17 % of the world production of petrochemicals in 2015 and due to many supporting circumstances, such as the availability of feedstock, experience and funds to expand. It would produce well more than a third of the world production of petrochemicals in twenty years from now. The main producers of petrochemical and chemical output are Saudi Arabia 50%, Iran 27%, Qatar 9%, Kuwait 5%, Oman 5%, United Arab Emirates 3% and finally Bahrain 1%. By category, the main petrochemical and chemical output are basic petrochemicals 37%, fertilizers 28%, polymers 20% and intermediates 15%.

Table 1-2: The growth and share in the World production capacity of petrochemicals (million ton/year) in the Middle East between 2009 and 2015.

Product	World 2009	ME 2009	ME share %	World 2015	ME 2009	ME share %
Ethylene	133.5	16.8	12.6	156.5	32.0	20.5
PE	83.0	10.7	12.9	113.0	20.3	18.0
MEG	23.3	6.0	26.0	37.9	10.8	28.5
Propylene	87.5	5.1	5.8	105.0	10.1	9.6
PP	54.4	5.05	10.8	73.0	9.5	10.8
Total	381.7	43.65	11.44	485.4	82.7	17.04

Source: GPCA.

8

Chapter Two
Petrochemicals from Synthesis Gas

SYNTHESIS GAS 10
AMMONIA 14
UREA...17
METHYLAMINES...17
NITRIC ACID 18
MELAMINE 19
METHANOL...20
FORMALDEHYDE..22
FORMIC ACID..23
GAS TO LIQUIDS 24
HYDROFORMYLATION OF OLEFINS 27
OXO PRODUCTS 30
OLEFIN CARBONYLATION 32

Synthesis gas

Synthesis gas is the common name for a mixture of hydrogen gas and carbon monoxide in different proportions. It is an excellent basis for the synthesis of some valuable petrochemicals such as ammonia, methyl alcohol, formic acid, acetic acid, formaldehyde, phosgene and others. This mixture has recently had an excellent role in the synthesis of petroleum liquids, as will appear in this chapter, particularly with the drastic increase in oil prices, using modern techniques that are simply new developments of the old Fischer-Tropsch synthesis. Figure (2-1) shows the main petrochemical derivatives of synthesis gas.

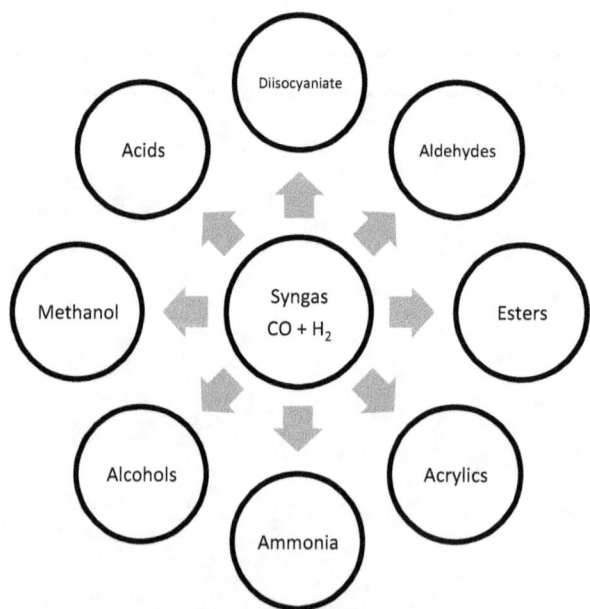

Figure 2-1: The main petrochemical derivatives of synthesis gas.

The basis for the manufacture of synthesis gas at the beginning of the nineteenth century was the interaction of coal with water vapor at high temperatures:

$$C + H_2O \longrightarrow CO + H_2$$

Then the era of oil came, where the synthesis gas industry became prosperous and popular to produce hydrogen, which through interaction with nitrogen gas yields ammonia. The latter is badly needed for the production of nitrogen fertilizers:

$$N_2 + 3H_2 \longrightarrow 2NH_3$$

Hydrogen was obtained at the beginning of the twentieth century from the cracking of natural gas, but this method was replaced quickly due to lack of ability of economic competition:

$$CH_4 \longrightarrow 2H_2 + C$$

Natural gas was reacted with steam to get synthesis gas with the required fantastic advantage of a high proportion of hydrogen gas:

$$CH_4 + H_2O \longrightarrow CO + 3H_2$$

This method is called steam reforming in which nickel is used as a catalyst at temperatures ranging from 800 to 850°C and a pressure ranging between 25 and 40 atmospheres; however, the feed must be sulfur-free to avoid poisoning the catalyst. The steam to methane ratio has be 3:1 to be economic and to reduce the carbon deposition on the catalyst. The same method used oil as a feedstock with little modification. A small amount of carbon dioxide is produced due to reaction of steam with carbon monoxide as seen in the following equation:

$$CO + H_2O \longrightarrow CO_2 + H_2$$

Another method to prepare synthesis gas from natural gas is called the partial oxidation of methane as follows:

$$CH_4 + \tfrac{1}{2} O_2 \longrightarrow CO + 2 H_2$$

In this reaction, oxygen is used, not air, as the nitrogen in the air will dilute the reactants. The method could be applied to the partial oxidation of oil hydrocarbons, from methane until fuel oil, at temperatures of more than 1500°C and a pressure range between 130 and 140 atmospheres. Carbon dioxide is formed in this reaction by the interaction of methane with oxygen quickly as follows:

$$CH_4 + 2 O_2 \longrightarrow CO_2 + 2 H_2O$$

Some carbon dioxide and water formed will react slowly with methane giving more carbon monoxide and hydrogen as follows:

$$CH_4 + CO_2 \longrightarrow 2 CO + 2 H_2$$

$$CH_4 + H_2O \longrightarrow CO + 3 H_2$$

Table (2-1) summarizes the formation of carbon dioxide as a by-product in the preparation of synthesis gas through steam reforming and partial oxidation.

It can be noted from Table (2-2) that the ratio of carbon monoxide to hydrogen varies from an application to another and the ratios need to be adjusted as defined in the industrial applications. The different methods to control the ratios will be summarized in the following paragraphs.

Carbon dioxide is separated from the mixture by solvent extraction in suitable solvents such as monoethanolamine and diethanolamine. Table (2-2) summarizes the most prominent industrial uses of synthesis gas.

Table 2-1: Formation of carbon dioxide as a by-product in the preparation of synthesis gas.

Yield % using methane	H_2	CO	CO_2
By steam reforming	10	15	75
By partial oxidation	5	45	50

Table 2-2: The most prominent industrial uses of synthesis gas.

Mixture	Industrial application
H_2	Hydrocracking and hydrogen treatment in refineries
$N_2 + 3H_2$	Ammonia by Haber-Bosch Process
$CO + H_2$	Synthesis of aldehydes and alcohols
$CO + 2H_2$	Synthesis of methanol
$CO + H_2O$	Synthesis of formic and acetic acids
$CO + 2H_2$	Fischer-Tropsch olefin synthesis
$CO + 3H_2$	Fischer-Tropsch hydrocarbons synthesis

To reduce the content of carbon monoxide in synthesis gas, the latter is mixed with steam and passed over a catalyst. Carbon monoxide is converted into carbon dioxide. Steam turns, after losing the oxygen atom, into hydrogen and thus increases the proportion of hydrogen and reduces carbon monoxide:

$$CO + H_2O \longrightarrow CO_2 + H_2$$

When the increase in carbon monoxide gas is necessary, for the manufacture of a certain synthesis process, additional carbon dioxide is added to react with methane during the steam reforming to increase carbon monoxide, as illustrated by the following equation:

$$CH_4 + CO_2 \longrightarrow 2\,CO + 2\,H_2$$

13

The former reactions are used to modify the ratios of hydrogen and carbon monoxide to each other. In the oil industry and the manufacture of ammonia, where pure hydrogen is required, carbon monoxide is separated from hydrogen or reacted chemically as its presence poisons the catalysts used.

Although naphtha is often used for the production of synthesis gas, the use of natural gas is characterized by three advantages:

a) Naphtha has to be gasified before reaction, which means additional equipment, energy and financial costs.

b) The sulfur content in naphtha is high and its removal is more difficult than the removal of hydrogen sulfide in natural gas, for example. Therefore, naphtha has to be treated carefully and costly to remove sulfur before use.

c) A larger quantity of carbon dioxide is formed when using naphtha, and the disposal costs are greater:

$$Naphtha\ (CH_{2.1}) \quad + 2\ H_2O \longrightarrow \quad CO_2 + 3.05\ H_2$$

$$CH_4 + H_2O \longrightarrow CO + 3\ H_2$$

Ammonia

There are several ways for the synthesis of ammonia due to its utmost importance as one of the basic petrochemical feedstocks, applied in the manufacture of nitrogen fertilizers, and in the preparation of a large number of intermediates and final petrochemicals.

There are five companies producing large complexes for ammonia industries, namely: Kellogg, Linde, Synetix, Krupp Uhde and H.

Topsoe. The production of ammonia ranged between six hundred and eighteen hundred metric tons per day.

The modern ammonia industry begins from hydrocarbons to obtain hydrogen through steam reforming. Different hydrocarbons are used starting from natural gas, liquefied petroleum gas (LPG), liquid natural gas (LNG), heavy naphtha, and ending with heavy diesel and fuel oil. Often, many companies use a range of suitable materials as feedstock of steam reforming for hydrogen production. According to the following reaction, beginning with methane as an example for the production of synthesis gas, a mixture of various compositions of hydrogen and carbon monoxide is obtained:

$$CH_4 + H_2O \longrightarrow CO + 3\,H_2$$

The manufacturing companies are competing to lower the temperature and pressure of the steam reforming process required for the production. The most widely accepted processes in the world of industry are the easily replicated, least costly and most efficient. Usually the process of removing the sulfur from the heavy hydrocarbons starts before, as well as in the case of using natural gas feedstock that contains high sulfur content.

Carbon monoxide can be disposed of through conversion into carbon dioxide by air oxidation:

$$2\,CO + O_2 \longrightarrow 2CO_2$$

Carbon dioxide is absorbed in monoethanolamine or diethanolamine as follows:

$$CO_2 + H_2O + CH_3N(CH_2CH_2OH)_2 \longrightarrow CH_3{}^+N(CH_2CH_2OH)_2\ HCO_3^-$$

There are many ways of getting rid of harmful acid gases in the petrochemical industry. Potassium carbonate is used sometimes in absorption of carbon dioxide:

$$CO_2 + K_2CO_3 + H_2O \longrightarrow 2\,KHCO_3$$

The disposal of the little remnants of carbon dioxide is accomplished by converting it into methanol to keep the hydrogen pure, to satisfy the requirements of the ammonia synthesis:

$$CO_2 + 3\,H_2 \longrightarrow CH_3OH + H_2O$$

Nitrogen required for the production of ammonia is drawn from the air:

$$N_2 + 3\,H_2 \longrightarrow 2\,NH_3$$

The resulting ammonia is condensed by cooling at -33°C. The manufacturing companies retain secrets of the catalysts used that make the efficient conversion possible in the production of ammonia and steam reforming.

Kellogg's method is characterized by the manufacture of ammonia that costs 10% less than the traditional method as well as high efficiency and energy saving. This company has built hundreds of complexes for the manufacture of ammonia at different locations in the world. Table (2-3) summarizes the most famous modern international ammonia technologies.

Table 2-3: The most famous international ammonia technologies.

Company	Feedstock	Production Ton/day	Energy Gcal/ton
H. Topsoe	Natural gas, heavy naphtha		
Krupp Uhde	Gas, naphtha, coal, crude oil, residuals, methanol	500-1800	6.65
Linde	Hydrocarbons	230-1500	7
Synetix	Hydrocarbons	1000-1750	6.75
Kellogg	Hydrocarbons	1000-1850	

Urea

The synthesis of urea is accomplished in two steps. In the first step, ammonia reacts with carbon dioxide to give ammonium carbamate, at 180 to 215°C, under a pressure ranging between 200 and 250 atmospheres. Ammonium carbamate is deprived of water in the second step, to yield urea in 60 to 70%:

$$2\ NH_3\ +\ CO_2\ \longrightarrow\ NH_4CO_2NH_2$$

$$NH_4CO_2NH_2\ \longrightarrow\ NH_2CONH_2\ +\ H_2O$$

Methylamines

In this process, mono-, di- and trimethylamines are prepared from dry ammonia and methanol according to the following equations:

$$NH_3\ +\ CH_3OH\ \longrightarrow\ H_2NCH_3$$

$$H_2NCH_3\ +\ CH_3OH\ \longrightarrow\ HN(CH_3)_2$$

$$HN(NH_3)_2\ +\ CH_3OH\ \longrightarrow\ N(CH_3)_3$$

Ammonia is pumped with methanol continuously, together with the recycled liquids and gases in calculated rates and under full control over an evaporator and heat exchanger and then pumped hot (Super heater) to the catalytic converter charged with the amination catalyst. Most of the heat used is generated from the exothermic reaction in this heat cultivating system. Raw materials and products are pumped to the four distillation columns where excess ammonia is separated and a part of the isotropic trimethanolamine is separated and recycled. The remaining portion goes where water is added to the

17

extractive distillation, and pure trimethanolamine is extracted for ultimate storage or recycling. Residues of monomethanolamine and dimethanolamine are separated. The leftovers are pumped for separation of ammonia and methanol from the top to be recycled, and the water is discharged from the bottom of the final treatment. The yields obtained in this process exceed 99.6%. Upon installation of a fifth column, it becomes easy to raise the efficiency of production and recycling, and the catalyst life becomes 24-36 months.

This technology is used in tens of countries in the World, and is owned by Acid Amine Technologies.

Nitric acid

Nitric acid is one of the most notable products of ammonia. It reacts with ammonia to give ammonium nitrate, the famous nitrogen fertilizer, in addition to its importance in the production of industrial explosives, adipic acid, nitrobenzene, and toluene diisocyanate, among others.

Nitric acid is synthesized by oxidation of ammonia, by passing a mixture of 10% ammonia in the air on a network of platinum and rhodium as a catalyst, at high temperature to give nitrogen oxide, which in turn is oxidized in the air, to give nitrogen dioxide. Nitrogen dioxide is absorbed in water to produce nitric acid. The following equations summarize the synthesis of nitric acid from ammonia:

$$4\,NH_3 + 5\,O_2 \longrightarrow 4\,NO + 6\,H_2O$$

$$2\,NO + O_2 \longrightarrow 2\,NO_2$$

$$3\,NO_2 + H_2O \longrightarrow 2HNO_3 + NO$$

Melamine

Melamine is manufactured from urea in two steps. Cyanic acid and ammonia are obtained in the first step. In the second step cyanic acid molecules react to yield melamine and carbon dioxide, at about 400°C. Melamine is sublimed, and then extinguished in water, and the suspended melamine is separated by centrifuge and dried:

$$H_2NCONH_2 \longrightarrow HNCO + NH_3$$

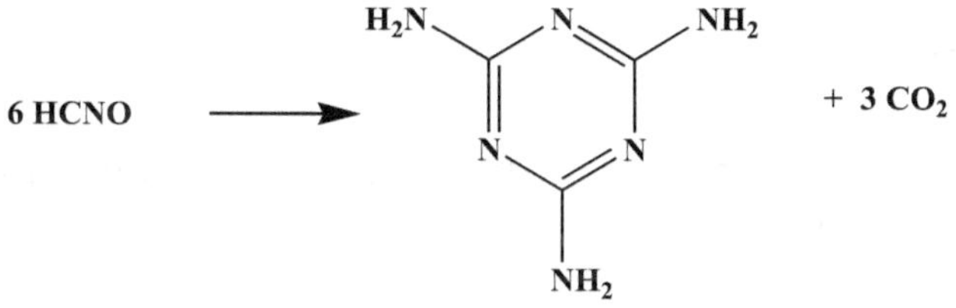

6 HCNO $\longrightarrow$ + 3 CO$_2$

Carbon dioxide and ammonia produced in this industrial process can be recycled and reacted for the preparation of more urea. This encourages the existence of multiple plants in the industrial complex.

Melamine can be synthesized starting from cyanimine, which is produced in turn from calcium carbide in several steps as shown in the following equations:

$$CaO + 3C \xrightarrow{2000} CaC_2 + CO$$

$$CaC_2 + N_2 \xrightarrow{1000} CaCN_2 + C$$

$$CaCN_2 + H_2SO_4 \longrightarrow CaSO_4 + H_2N\text{-}CN$$

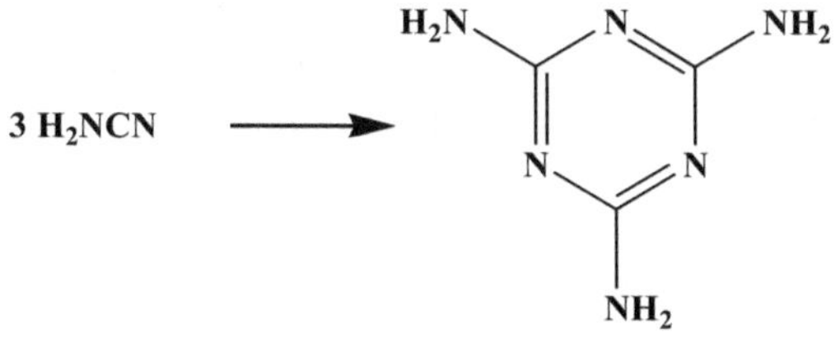

3 H$_2$NCN $\longrightarrow$

Melamine has paramount importance in the synthesis of melamine-formaldehyde resins, which are characterized by solidity, water resistance, ability of formation, colorability, and heat resistance. It is used for the manufacture of kitchenware including dishes, spoons, etc. Melamine-formaldehyde kitchenware sets had spread recently strongly, especially in the homes of the poor. The plastics are obtained by condensing melamine with formaldehyde to give strong cohesive networks of plastics.

Methanol

Methanol is synthesized from natural gas or associated gas through steam reforming in a step or two to produce synthesis gas as seen in the following equations:

$$CH_4 + \tfrac{1}{2} O_2 \longrightarrow CO + 2 H_2$$

Methanol is synthesized from synthesis gas as follows:

$$CO + 2H_2 \longrightarrow CH_3OH$$

There are five natural steps to complete the normal process of methanol synthesis. The first step is purifying the raw materials, disposal of harmful substances and the acidic sulfur compounds. The second is the steam reforming and/or partial oxidation of natural gas. The third step is setting the ratios required in the synthesis gas for methanol synthesis. The fourth step is the synthesis of methanol by reacting carbon monoxide with hydrogen. The fifth step is the distillation of methanol for purification. The feedstock used in the preparation of synthesis gas ranges between natural gas, associated gas, liquefied petroleum gas, LPG, oil and coal, and even includes all the different gases and liquids of the petrochemical waste.

The modern methanol industry plants are characterized by the possibility of conversion for the production of ammonia. This

20

excellent feature of switching from production of ammonia to methanol and vice versa allows flexibility in the work and freedom to keep up with the production demands in the World markets.

Table 2-4: The most notable characteristics and differences between the most famous companies and technologies for methanol production.

Comp.	Reforming	Characteristics	Energy Gcal/t	1000 ton/day	Feed
Haldor Topsoe	2 steps	- Convertible to ammonia - 10% cheaper	7	3-10	NG
Kvaerner	2 steps	-	5.7	5-7	NG
Krupp Uhde	Steam reforming	- Gas or naphtha feed	5.7		NG or naphtha
Lurgi		-	7	Up to 10	NG
Synetix		- Gas to coal feed		3 or more	NG, naphtha, coal etc.

These magnificent factories have the ability of production of large amounts of methanol, ranging between three and ten thousand metric tons per day. Five giant international companies are competing on the technologies of the methanol industry and the establishment of their complexes. Figure (2-2) summarizes the most prominent chemical derivatives of methanol. Table (2-4) summarizes the most notable characteristics and differences between these five companies and technologies.

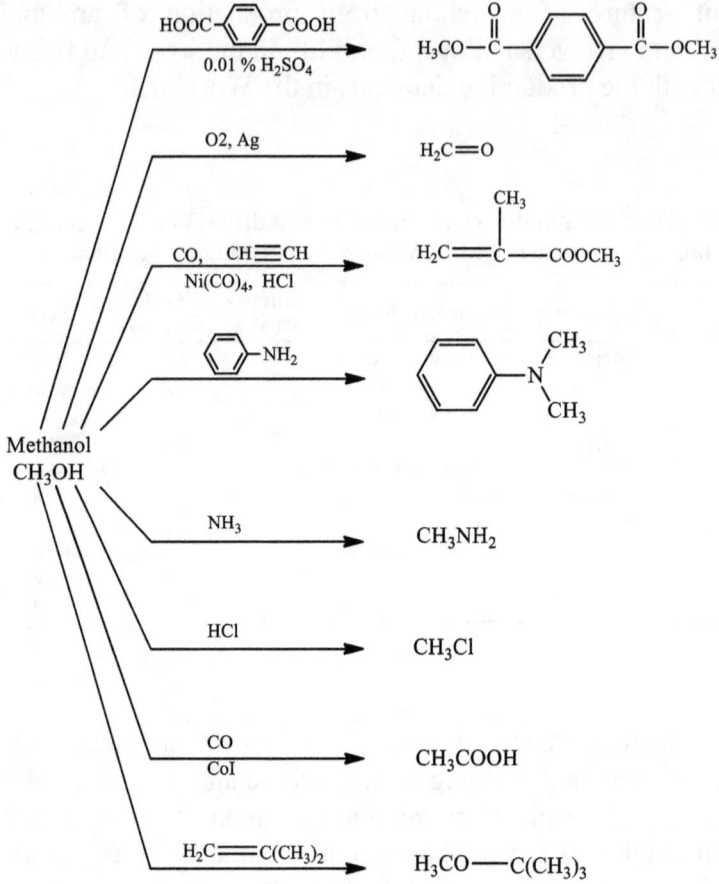

Figure 2-2: The main petrochemical derivatives of methanol.

Formaldehyde

The synthesis of formaldehyde is accomplished by the oxidation of methanol by oxygen at 250°C in the presence of molybdenum oxide as a catalyst activated with iron oxide according to the following equation:

$$CH_3OH + 1/2\,O_2 \longrightarrow CH_2=O + H_2O$$

The product can be absorbed in water to give the required concentration of (37%), called formalin, or is reacted with urea to yield urea-formaldehyde resin. The main principal uses of formaldehyde are in the production of urea-formaldehyde, phenol-formaldehyde, acetal resins, melamine-formaldehyde, and several other chemicals, such as butane dialcohol, pentaerythritol, hexamethylenetetraamine, isoprene and others.

Formic acid

Formic acid and its esters are synthesized from carbon monoxide using a basic catalytic such as sodium hydroxide, calcium hydroxide or sodium methoxide, at 110-150°C, and 8-30 atmospheric pressure, to prepare formic acid, and at 70°C, and 20-200 atmospheric pressures for the synthesis of formic acid esters:

$$CO + H_2O \longrightarrow HCOOH$$

$$CO + ROH \longrightarrow HCOOR$$

It is worth mentioning that formamide and its N-methyl derivatives are among the best polar organic solvents used in the chemical industry in massive quantities, as in the extraction of acetylene, and other materials:

$$HCOOCH_3 + HN(CH_3)_2 \longrightarrow HCON(CH_3)_2 + CH_3OH$$

The synthesis of di-N-methyl formamide directly from carbon monoxide and dimethylamine in the presence of sodium methoxide catalyst proceeds as follows:

$$CO + HN(CH_3)_2 \longrightarrow HCON(CH_3)_2$$

23

Gas to liquids

Many exploration and production companies discovered vast amounts of natural gas in the past decades during their search for crude oil, but they were not interested in it; they were concerned with crude oil alone. Large quantities of natural gas were burned in the oil fields in the past decades. The crude oil producing companies wish to develop the newly discovered gas fields that fall under their hands, especially those fields that do not seem economically feasible, located far away from the consumer markets and nearby locations, or which can not be disposed of by burning, or re-injection into oil stocks.

Some remote fields do not find an appropriate use, and pumping gas to a pipeline becomes costly and impractical. The modern environmental regulations prohibit the disposal of these gases, associated with crude oil, through burning. The re-injection of gas is sometimes costly, and in many times the gas cannot be injected without damaging the field, and sometimes the restoration of the injected gas again is unlikely.

About 60% of the gas fields in the world suffer from problems of this kind. The owners of these fields find difficulties and inability, in whole or part, in the development of their fields and benefit from their stockpiles.

Despite some real obstacles in this exploitation, it represents an acceptable way for the gas-producing nations, the global oil companies, and everyone is competing to try to develop and maintain the technical means and the promising future, to produce the highly demanded liquid petroleum.

The bright idea of GTL starts from exploiting distant and remote areas gas fields to convert natural gas into carbon monoxide and hydrogen first, which is known as synthesis gas, then convert the latter through the well-known Fischer-Tropsch process to the petroleum liquids desired, as shown in Figure (2-3).

24

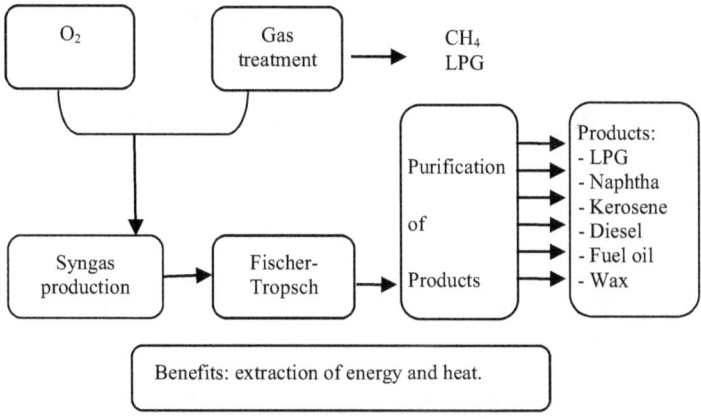

Figure 2-3: The gas-to-liquids process.

Although GTL process is complex, and requires substantial capital, it has similarities in the global market, such as the production of ammonia and methanol that rely on natural gas, synthesis gas and technologies similar to the old Fischer-Tropsch method. GTL process is struggling to sail in the midst of the world petrochemicals, trying to overcome the enormous difficulties facing it, such as the size of economic production, energy efficiency, design of reactor, use of raw materials and products reward. Although attempts are at the beginning, the idea will be delightful and profitable in the future.

The four steps of GTL process are as follows:

1) The purification of natural gas to get rid of the pollutants poisoning the catalysts such as sulfur compounds. Praise be to Allah that the Yemeni natural gas is entirely free of those contaminants. In this process, propane and butane gases are also separated for sale as liquefied petroleum gas. For propane and butane gas, many uses are known. Ethane is separated, and cracked for the production of ethylene gas that is used in the manufacture of the different polyethylene types. Methane remains alone for the following GTL

operations. It is clear that the process has an optimal and most comprehensive use of natural gas.

2) The production of synthesis gas by reacting methane with air and steam through steam reforming technology or partial oxidation technology, or both; the so-called auto thermal reforming.

3) Fischer-Tropsch process, where synthesis gas is converted into long chain paraffins, light olefins, high molecular weight waxes and water.

4) The treatment and improvement of liquid hydrocarbons, where the final desired products are treated such as low sulfur diesel, naphtha, waxes, base oils for lubrication oil and grease. These GTL products are usually entirely free of sulfur and aromatics, and these qualities are favored and highly demanded as one of the requirements for anticipated future specifications of some remarkable expensive petroleum fractions. Unfortunately, these specifications are not available currently without the high cost of production. Hence, one of the best applications of GTL products is as a blend feedstock in refineries to improve qualities and specifications.

GTL is a chemical process that converts natural gas into ultra clean liquid fluids. It needs reasonably cheap natural gas feedstock that is guaranteed for a long period of time, to cover the high cost of construction. This requirement is available in the fields of particularly the Middle East, Russia, Africa and South America. Thus, it can be seen that cheap natural gas is found at locations where it is difficult to establish these projects. GTL projects are blossoming around the world. The number of Gas-To-Liquids (GTL) feasibility studies being carried out around the World continues to increase. Many of these projects, of course, are under consideration within corporations on a confidential basis and we will not hear about them for some time, if ever. Table (2-5) summarizes the most prestigious projects announced to convert gas to liquids in the world with its productivity in 2002. Many other GTL projects were announced after that with bigger capacities all over the world. The typical range of products in the GTL process is summarized in Table (2-6).

Table 2-5: The announced GTL projects (2002) and their capacities.

Location	No. of plants	Capacity bbl./day
Australia	3	115000
Argentina	1	75000
Egypt	2	145000
Indonesia	1	70000
Iran	1	70000
Nigeria	1	30000
Qatar	3	290000
Trinidad	1	75000
Total	**13**	**870000**

Table 2-6: The typical range of products in a GTL process.

Product	Weight%	Market end-use
C_1-C_4 (gas, LPG)	5-10%	Petrochemicals feedstock
C_5-C_9 (naphtha)	15-20%	Petrochemicals and gasoline via reforming
C_{10}-C_{16} (kerosene)	20-30%	Jet fuel
C_{17}-C_{22} (diesel)	10-15%	Diesel fuel
C_{22+}	30-45%	Waxes, lubricants

Hydroformylation of olefins

Hydroformylation or oxo synthesis is an industrial process for producing aldehydes from olefins, carbon monoxide and hydrogen. In this process, the chain length of the olefin increases by one carbon. The main products of modern oxo synthesis are normal butane and 2-ethyl hexanol, which is used in the manufacture of plasticizers.

27

Oxo synthesis includes the synthesis of a broad spectrum of olefins from C_2 to C_{20}, straight or branched, whether the double bond is internal or terminal. The yields of this process are mixtures of normal and iso-aldehydes that prefer to give a high proportion of normal aldehydes compared to the iso-aldehydes:

$$RCH=CH_2 + CO + H_2 \longrightarrow RCH_2CH_2CHO + \overset{\overset{\textstyle CH_3}{|}}{R}CHCHO$$

Of course, the structure of the olefin has an impact on the yield, and the structure of the product remains as it is in the aldehyde and does not isomerize. The double bonds might isomerize if they are internal, and this will lead to mixtures of aldehydes.

The double bonds might isomerize and usually migrate to the terminals where the π-π complexes are more stable. Alpha olefins react directly compared with those suffering from steric hindrance in the case of interior olefins.

The catalysts of this process depend on the central atoms of cobalt (Co), rhenium (Rh) or ruthenium (Ru), where the ligand is an alkyl or aryl group. In the industry, the use of cobalt is preferable as a metal or compound, but in its active state:

$$2\ HCo(CO)_4 \longrightarrow Co_2(CO)_8 + H_2$$

Normal aldehydes are preferable. Increasing their ratio, compared to iso-aldehydes is possible by controlling the reaction conditions of pressure, temperature, partial pressure of carbon monoxide, and selecting the appropriate catalytic ligands and the central atom. The terminal complex, where the C-C is sterically hindered, gives a larger percentage of the normal aldehyde. It is thus clear that voluminous ligands increase the proportion of normal aldehydes. However, the reaction yields become lower and the reaction speed is reduced. At the same time the hydrogenation activities increase where the olefins are hydrogenated to paraffins and the aldehydes into alcohols. With regard to the central atom, rhenium instead of cobalt reduces the proportion of normal aldehydes but increases the proportion of the branched olefins. However, the rhenium with phosphine ligands lead to the highest possible proportion of normal aldehydes. One of the most notable side reactions of hydroformylation is the hydrogenation of the alkenes to alkanes and the isomerization of internal alkenes to terminal alkenes.

Among the secondary reactions in this process are the hydrogenation of aldehydes formed to the corresponding alcohols, the Aldol condensation between the aldehydes themselves, and the reaction of aldehydes by hydroformylation giving formic acid esters and, finally, the formation of the acetals from aldehydes and alcohols.

The following ratios were obtained (weight %) from the following typical reaction:

$$RCH=CH_2 + CO + H_2 \longrightarrow RCH_2CH_2CHO + \overset{\overset{\textstyle CH_3}{\textstyle |}}{R}CHCHO$$

C$_4$ aldehydes	80
Alcohols and butyl formates	10-14
Other	6-10
Normal to iso-aldehydes	25:75 to 20:80

A number of variables would affect the rate of formation of aldehydes and their structures in the oxo reactions such as the low partial pressure of carbon monoxide. Hydroformylation reaction prefers the increase of partial pressure of hydrogen to increase the conversion of olefins, but this reduces the selectivity and directs to the formation of alkanes and alcohols as side products.

Rhenium phosphine catalysts are characterized by their higher percentages of normal aldehydes, decrease in partial pressure, as well as ease of operation. The reaction mechanism occurs in four steps according to the following equations:

1) HRhCO(PPh$_3$)$_3$ forms coordination sites through the disintegration of the ligand.

2) Rhenium compounds formed are added to the resulting olefins to give π-compounds.

$$HRhCO(PPh_3)_3 + CH_2=CHCH_3 \longrightarrow HRhCO(PPh_3)_2 \uparrow CH_2=CHCH_3$$

3) Re-coordination of π-sigma to form Rh-CO compounds.

$$\longrightarrow C_3H_7RhCO(PPh_3)_2$$

4) Introduction of CO to the acyl complex.

$$\xrightarrow{+ \ CO} C_3H_7(CO)RhCO(PPH_3)_2$$

5) Hydrogenation of Rh-CO bond.

$$\xrightarrow{+ \ H_2 + PPh_3} C_3H_7CHO + HRhCO(PPh_3)_3$$

Oxo products

Aldehydes are the primary products of hydroformylation; this track is known as the oxo reaction, and all the industrial aldehyde products produced are called oxo products.

Although the aldehydes themselves have no real importance as final products, they are active intermediates and famous for the production of the oxo alcohols, carboxylic acids, and Aldol condensation products. The oxo aldehydes are converted to primary amines in limited quantities, as follows:

$$RCHO + NH_3 \longrightarrow RCH_2NH_2 + H_2O$$

The aldehydes, Aldol condensation products and enals produced are hydrogenated from oxo aldehydes to alcohols. The term oxo alcohols refer in particular to the three principal products that depend on the oxo reaction; namely, butanol, isobutanol, and 2-ethyl hexanol.

Oxo alcohols are produced from oxo aldehydes through hydrogenation, catalyzed by nickel or copper in the gas phase, and in the presence of a nickel catalyst in the liquid state:

$$RCH_2CH_2CHO + H_2 \longrightarrow RCH_2CH_2CH_2OH$$

The use of oxo alcohols relies on their length. C_4-C_6 alcohols are used as is, or in the form of esters, as solvents. C_8-C_{13} alcohols are used as esters such as phthalates. They are highly demanded in the marketplace as plasticizers. C_{12}-C_{19} alcohols are used, after conversion to sulfonates, as raw materials in the manufacture of synthetic detergent (RCH_2SO_2OH), or the fabrics industry.

Oxo carboxylic acids are obtained from the reaction of oxo aldehydes with oxygen in the presence of a homogeneous catalyst or without a catalyst in the liquid phase, via the carboxylic acid peroxide, in the presence of a mineral salts catalyst:

$$CH_3CH_2CHO \xrightarrow{O_2} CH_3CH_2COOH$$

31

Oxo carboxylic acids are used in the form of esters as solvents, in the form of acids to formulate alkyd resins, in the form of salts for speeding up drying in the dyes industry, and in the form of vinyl esters in the production of dispersion copolymers.

The products of Aldol condensation from aldehyde compounds have been dealt with elsewhere in this book.

Olefin carbonylation

The reaction of olefins with carbon monoxide and a nucleophile in the presence of a metal carbonyl catalyst will lead to the formation of a carboxylic acid and its derivatives. This process is called Reppe carbonylation:

$$CH_2=CH_2 + CO + HX \longrightarrow CH_3CH_2COX$$

$$x = -OH, -OR, -SR, -NHR, ..etc.$$

The carboxylation by insertion of a carboxyl group (COOH) in the olefin is called hydrocarboxylation.

Koch and Reppe hydrocarboxylation differ in three ways: the catalyst, reaction conditions and products. The hydrocarboxylation in Reppe reaction is catalyzed by metals carbonyls, preferably $Ni(CO)_4$, homogeneous in the liquid state, at high temperature and pressure. The production of propionic acid from ethylene, water and carbon monoxide is given as an example:

$$CH_2=CH_2 + CO + H_2O \longrightarrow CH_3CH_2COOH$$

Sodium and calcium salts of propionic acid are used in conservation and the esters are used as solvents and plasticizers, such as glycerin tripropionate and phenyl esters as an important comonomer. It is the same raw material required in the manufacture of insecticides.

Koch carbonylation is used particularly in the synthesis of tertiary carboxylic acids from olefins. Unlike Reppe carbonylation, proton catalysis leads to the formation of a double bond, as well as isomerization of the structure. The basis for Koch reaction is the formation of a more stable carbonium ion through isomerization, the addition of carbon monoxide to the acylium cation, and finally the reaction with water or alcohol to give the required acid or ester, respectively:

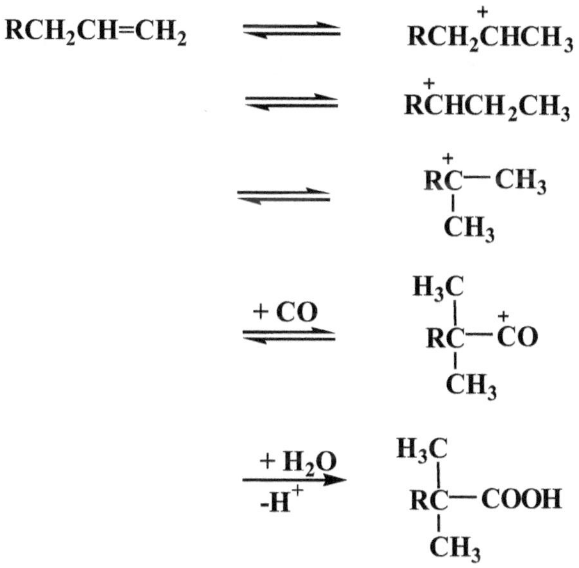

$$RCH_2CH{=}CH_2 \rightleftharpoons RCH_2\overset{+}{C}HCH_3$$

$$\rightleftharpoons R\overset{+}{C}HCH_2CH_3$$

$$\rightleftharpoons \underset{\underset{CH_3}{|}}{R\overset{+}{C}{-}CH_3}$$

$$\overset{+CO}{\rightleftharpoons} \underset{\underset{CH_3}{|}}{\overset{\overset{H_3C}{|}}{R C{-}\overset{+}{C}O}}$$

$$\xrightarrow[-H^+]{+H_2O} \underset{\underset{CH_3}{|}}{\overset{\overset{H_3C}{|}}{R C{-}COOH}}$$

The following reaction is a typical example:

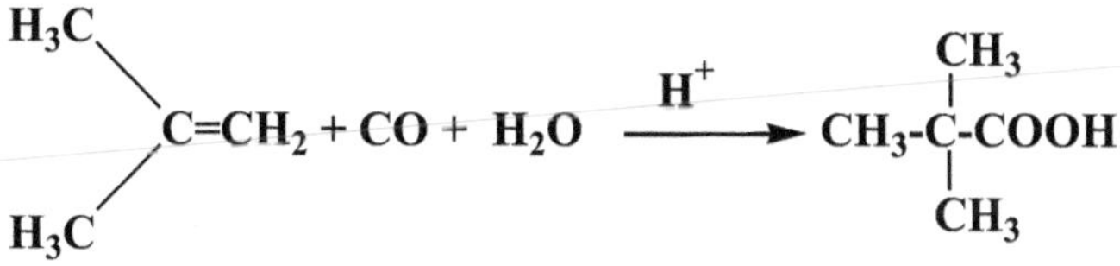

$$\underset{H_3C}{\overset{H_3C}{}}{\diagdown}{C}{=}CH_2 + CO + H_2O \xrightarrow{H^+} \underset{\underset{CH_3}{|}}{\overset{\overset{CH_3}{|}}{CH_3{-}C{-}COOH}}$$

Chapter Three
Petrochemicals from Ethylene

ETHYLENE 36

MECHANISM OF THERMAL CRACKING........................43

POLYETHYLENE..47

METALLOCENES 51

ETHYL BENZENE 53

STYRENE...55

POLYSTYRENE 58

LINEAR ALKYL BENZENE 59

LINEAR ALPHA OLEFINS 61

SABLIN NEW LINEAR ALPHA OLEFINS 61

ETHANOL (ETHYL ALCOHOL) 64

ACETALDEHYDE 66

ACETIC ACID 69

  Synthesis of acetic acid by oxidation of acetaldehyde................69

  Synthesis of acetic acid by oxidation of alkanes and alkenes......70

  Synthesis of acetic acid from methanol..................................71

  Uses of acetic acid 73

ETHYL ACETATE 74

ETHYLENE OXIDE 74

ETHYLENE GLYCOL 78

ETHYLENE CHLORIDE 79

ETHANOLAMINES 81

ETHYLENE BROMIDE 82

VINYL CHLORIDE 82

1,1,2-TRICHLOROETHYLENE 86

PERCHLOROETHYLENE 88

ETHYL CHLORIDE 89

VINYL FLUORIDE 90

TETRAFLOUROETHYLENE 91

VINYL ACETATE 91

Ethylene

Ethylene is the most fundamental primary petrochemical among all.

The world would consume, in the twenty first century, tens of millions of metric tons of ethylene annually, produced by hundreds of expertly manufactured factories characterized by wonderful performance. Most of the ethylene is consumed in the polyethylene industry in its different forms, which requires ethylene with extremely high purity, exceeding 99.95%. Four global institutions are competing to produce ethylene plants, or in fact, ethylene complexes because each one of them is a set of integrated plants and accessories. Based on each of these complexes, other minor secondary complexes consume what the first complexes produced and turn it into the so-called consumables, to dump the World with other high quality secondary products.

The respective complexes are based on the same principal idea of cracking ethane, or any other available hydrocarbons, ranging from liquefied petroleum gas (LPG), naphtha, diesel, fuel oil, or even crude oil itself. Table (3-1) summarizes the most prominent global uses of ethylene (ratio %) in 2000. Figure (3.1) illustrates the most prominent petrochemical uses of ethylene.

Olefins are unavailable in fossil fuels; they should be synthesized by cracking of petroleum fractions. Cracking can be classified in the following types:

1) Catalytic cracking

2) Hydrocracking

3) Thermal cracking

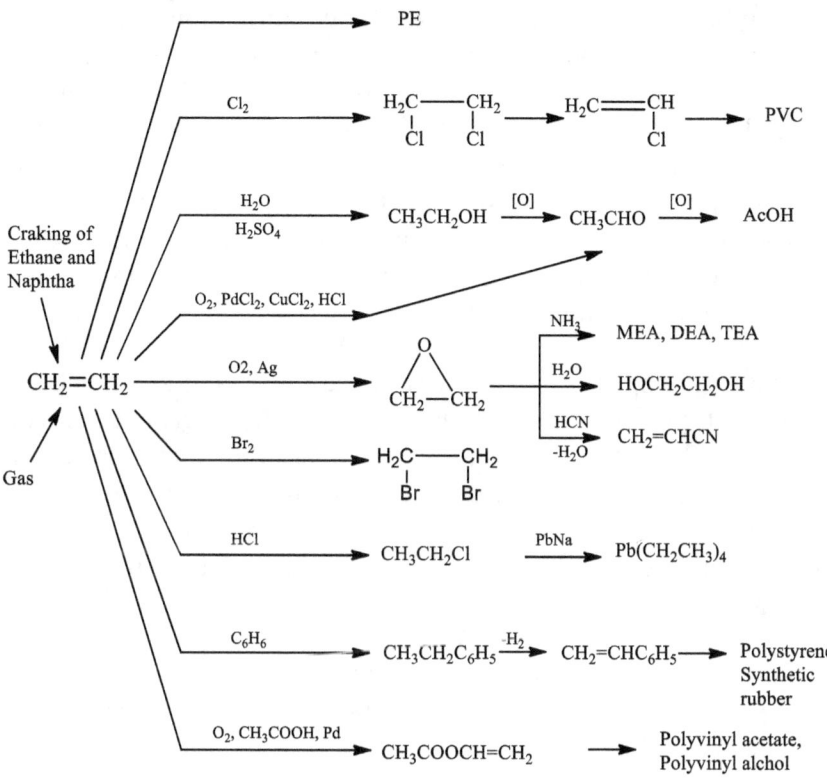

Figure 3-1: The most prominent petrochemical derivatives of ethylene.

Table 3-1: The most prominent global uses of ethylene (ratio %) in 2000.

Product	Consumption %
LDPE and HDPE	57
Vinyl chloride	14
Ethylene oxide and secondary products	13
Acetaldehyde and secondary products	1
Ethylbenzene and styrene	7
Other products: Ethylene, vinyl acetate, ethylene bromide, ethyl chloride, ethylamine, propionaldehyde	8
Total	**100**

The catalytic cracking is characterized by the formation of carbonium ions that react to give cyclic saturated naphthenic and aromatic compounds. The cracking catalyst is non-crystalline aluminum silicate that contains 10-15% of alumina, and lately increased to 25%. The newly used catalyst content ranges between 5 and 40% of the crystalline aluminum silicate (zeolite) with regular three-dimensional networks of aluminum oxide, tetrahedral silica, alternating with rare earth cations, for consistency and stability, in a mixture of non-crystalline silicate.

The hydrocracking process is characterized by using a bifunctional catalyst that performs hydrogenation and dehydrogenation at the same time. Saturated and branched hydrocarbons are preferably formed in an acid medium through carbonium ion reactions, such as dehydrogenation, isomerization and hydrogenation. Sulfur and nitrogen are removed (i.e. refining) during the process of hydrocracking by hydrogen.

The thermal cracking process is characterized by the absence of a catalyst, which proceeds via free radicals, and gives a high percentage of olefins.

The cracking process is conducted in two ways:

a) Initial reactions such as dehydrogenation, hydrogen exchange, carbon reactions, isomerization and cyclization.

b) Secondary reactions such as olefin polymerization, alkylation and aromatics condensation.

Olefin yields depend on three variables, and the relations between them, namely:

1. The temperature used. High temperatures tend to produce smaller olefins, such as ethylene and propylene, and reduce the ratio of higher olefins. The high temperature during the cracking process accelerates the reaction, raising the rate, and

this requires a shorter reaction time and/or a lower partial pressure.

2. The reaction time is influential as well, giving more secondary products if long, and more olefins when short.

3. The partial pressure of the reactants has a negative effect according to Le Chatelier's principle because of the generation of two or three moles of cracking products. Preferably, steam is used in reducing the vapor pressure, as in the case of steam cracking. The advantages of the reduced partial pressure are ease of condensation, and consequently separation, and low carbon breaking. Its significant disadvantage is the use of a big amount of energy in heating and cooling.

It can be concluded from the foregoing that the priority is the production of small olefins using high cracking temperatures, and short reaction times, which is the most efficient in the presence of steam. It is, therefore, necessary to follow one of two choices in cracking:

• Low temperature, less than 800°C, and a reaction time of one second,

• High temperature up to 900°C, and a reaction time of half a second,

• Partial hydrogenation of acetylene formed.

• And finally, water removal with ethanol.

The new industrial production of ethylene and propylene is through thermal cracking of natural gas, refinery gases, or crude oil fractions. The Yemeni diesel is sold to the European countries, usually Italy, for ethylene and propylene production, because of its very low sulfur content.

Even acidic oil fractions can be cracked in the presence of steam at higher temperatures that may reach up to 2000°C, in the presence of hydrogen and methane, to give a high yield of ethylene. Sulfur is hydrogenated into hydrogen sulfide, which is removed easily by absorption in basic solutions.

Ethylene and propylene can be obtained as polymerization purity grade of 99.95% through:

1) Selective hydrogenation:

$$HC \equiv CH \longrightarrow CH_2 = CH_2$$

$$H_3C - C \equiv CH \longrightarrow CH_3CH = CH_2$$

$$CH_2 = C = CH_2 \longrightarrow CH_3CH = CH_2$$

2) Extractive distillation with selective solvents such as dimethylformamide.

Table (3-2) summarizes the production ceiling of ethylene using different feedstocks.

The manufacturers are competing to provide plants that give greater yields of ethylene, lower production energy, cheaper prices for complexes, and reasonable operating costs.

Table 3-2: The production ceiling of ethylene using different feedstocks.

Feedstock Wt. %	Ethane	Propane	Naphtha	Light diesel	Vacuum diesel
Ethylene,	84	45	34.4	28.7	0.22
Propylene	1.4	14	14.4	14.8	12.1
Butadiene	1.4	2.0	4.9	4.9	5.0
Aromatics	4.0	3.5	14	14	8.5

Table 3-3: Comparison between the four main global pioneer companies in the manufacture of ethylene complexes in terms of yields obtained using various feedstocks as weight % and energy consumption per kg of ethylene produced.

Company	Yield Wt. %			Energy Kcal/kg ethylene		
	Ethane	Naphtha	Diesel	Ethane	Naphtha	Diesel
Lummus	84	35	22-35	3300	5000	
Kellogg	84	35	30			
Linde	83	35	25	3000	5000	6000
Stone & Webster	57		28	3000		6000

Table (3-3) shows a comparison between the four leading global pioneer companies in the manufacture of ethylene complexes in terms of yields obtained using various feedstocks as weight %, and the energy consumption per kg of ethylene produced.

Kellogg Brown Company has developed a fantastic technology for producing propylene and ethylene required in the global market from cheap materials and residues of oil fractions, C_4 to C_8, especially fractions C_4 and C_5 olefins that are available in the ethylene plants and refinery light gases such as catalytic cracking products, naphtha, and out of specifications gasoline. This process is called Superflex and it relies on a tremendous catalyst that does not require treatment to avoid poisoning materials such as sulfur compounds, water, oxygen or nitrogen. The catalytic conversion of paraffins and olefins to the required products is regenerated constantly and automatically, which allows the use of high temperature. Table (3-4) summarizes the ideal distribution for the production of ethylene and propylene using different feedstocks and technologies; namely, fluid catalytic crackers, Coker, and gasoline pyrolysis.

Figure (3-2) shows additions of ethylene complexes in the world by region between 1998 and 2003 with a total of 28 million tons. A clear development is noted in the Middle East, which comes after the

United States in expansion due to the availability of cheap gas in enormous quantities.

It can be seen from Table (3-4) that the quantity of materials heavier than propane in the thermal cracking products using gasoline feedstock amounts to about ten times as obtained using ethane. The heavier than propane materials are mixtures of butenes up to gas oil. It is a complex mixture that requires separation in a large number of distillation columns. The factories that use ethane feedstock require only a few separation columns. Therefore, the base cost of ethylene plants using naphtha as a starting material is about one and a half times the cost of the plant using ethane.

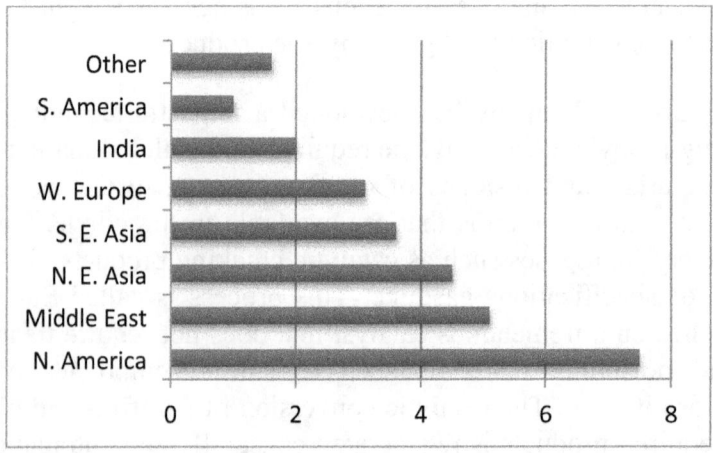

Figure 3-2: Additions of ethylene complexes in the World by region between 1998 and 2003 were 28 million tons.

Table 3-4: The ideal distribution for the production of ethylene and propylene using different feed stocks and technologies, fluid catalytic crackers, Coker, and gasoline pyrolysis.

Feed/Product	FCC Products	Coker	C$_4$ Pygas	C$_5$ Pygas
Diesel	6.13	6.11	2.7	12
Ethylene	20	8.19	5.22	1.22
Propylene	40	7.38	2.48	8.43
Propane	6.6	7	3.5	5.6
Gasoline C$_{6+}$	7.29	9.22	8.16	6.15

Mechanism of thermal cracking

It is known that the mechanism of reaction in the thermal cracking proceeds via free radicals. The reaction starts by breaking a C-C bond because it is weaker than the C-H bond:

$$CH_3\text{-}CH_3 \longrightarrow 2CH_3^\bullet$$

Then the methyl group radicals undergo all the typical reactions of free radicals, and they are many, such as an abstraction of hydrogen atoms as follows:

$$^\bullet CH_3 + CH_3^-CH_3 \longrightarrow CH_4 + CH_3\text{-}CH_2^\bullet$$

Ethylene can be obtained from the ethyl group radical:

$$CH_3^-CH_2^\bullet \longrightarrow CH_2{=}CH_2 + H^\bullet$$

$$H^\bullet + CH_3^-CH_3 \longrightarrow CH_3^-CH_2^\bullet + H_2$$

A large quantity of ethylene is produced in a series of the two previous reactions. The reactions come to an end when the free radicals react with each other:

$$2H^\bullet \longrightarrow H_2$$

$$H^\bullet + {}^\bullet CH_3 \longrightarrow CH_4$$

$$H + CH_3{}^-\overset{\bullet}{C}H_2 \longrightarrow CH_3{}^-CH_3$$

$$\overset{\bullet}{C}H_3 + CH_3{-}CH_2 \longrightarrow CH_3{-}CH_2{-}CH_3$$

$$2CH_3{-}\overset{\bullet}{C}H_2 \longrightarrow CH_3{-}CH_2{-}CH_2{-}CH_3$$

The presence of propylene and butenes is explained by the reaction of free radicals with ethylene as follows:

$$CH_3{}^\bullet + CH_2{=}CH_2 \longrightarrow CH_3CH_2CH_2{}^\bullet \longrightarrow CH_3CH{=}CH_2 + H^\bullet$$

$$CH_3CH_2{}^\bullet + CH_2{=}CH_2 \longrightarrow CH_3CH_2CH_2CH_2{}^\bullet \longrightarrow CH_3CH_2CH{=}CH_2 + \overset{\bullet}{H}$$

The mechanism of reaction of gasoline or naphtha cracking is more complex because of the presence of a mixture of compounds in them, although it follows the same mechanism in terms of the serial reaction of free radicals; however, it changes with the conditions of reaction such as temperature and pressure, as well as the change in the chemical composition of feedstock.

Taking n-octane as an example of thermal cracking in naphtha, the reaction will start with breaking a C-C bond:

$$CH_3(CH_2)_6CH_3 \longrightarrow CH_3(CH_2)_3\overset{\bullet}{C}H_2 + CH_3CH_2\overset{\bullet}{C}H_2$$

The free radicals formed would react by beta fission to give ethylene:

$$CH_3CH_2CH_2CH_2\overset{\bullet}{C}H_2 \longrightarrow CH_3CH_2CH_2{}^\bullet + CH_2{=}CH_2$$

$$CH_3CH_2CH_2{}^\bullet \longrightarrow CH_3{}^\bullet + CH_2{=}CH_2$$

The methyl free radical can abstract a hydrogen atom from another compound, to give a new free radical that might in turn split to give more ethylene:

$$CH_3^{\cdot} + C_8H_{18} \longrightarrow CH_4 + C_8H_{17}^{\cdot}$$

The chemical structure of naphtha affects the yield of ethylene. The branched compounds give a lower yield than the straight chain compounds. This can be seen in the cracking of 3-ethylheptane as an example:

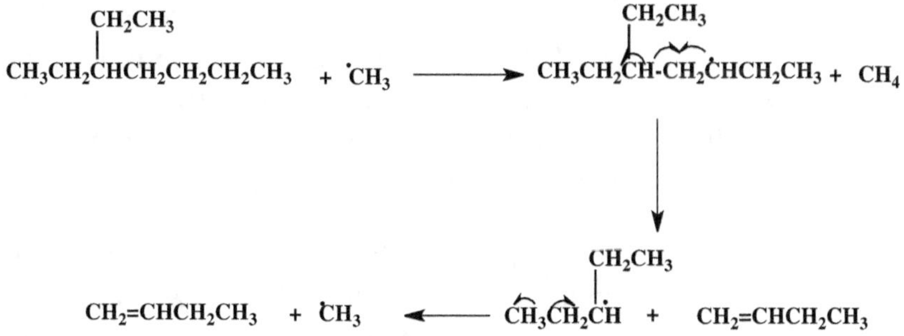

The same applies to the cyclic compounds:

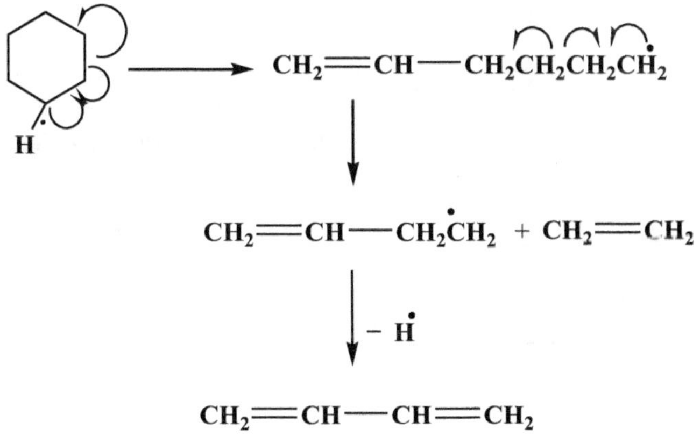

45

It is, therefore, clear that the quality of naphtha that is characterized by high proportions of straight chain alkanes gives greater revenues of ethylene.

The increase in pressure will give negative results because the purpose of thermal cracking is obtaining gases from liquid naphtha; hence, it is mixed with steam to reduce the partial pressure of hydrocarbons and also to prevent the deposition of coke on the pipelines of the oven. The coke reacts with steam as follows:

$$C + H_2O \longrightarrow CO + H_2$$

The high temperatures are desirable in the thermal cracking reactions to encourage the beginning of the reaction, the formation of free radicals, and help beta fission to form ethylene. In fact, the increase in temperature offsets an increase in the ethylene yields.

Table (3-5) summarizes the most beneficial technologies for steam cracking of ethane to ethylene, an easy option for the Arab States to produce cheap ethylene locally, without being subject to the international political and economic disturbances.

Table 3-5: The most powerful technologies for steam cracking of ethane.

Owner	Ethylene yield %	Kcal/kg of ethylene
ABB Lummus	84	3300
Kellogg Brown	84	-
Linde AG	83	3000
Stone & Web.	57	3000
Technip	83	3000

Polyethylene

The most valuable product obtained from ethylene is polyethylene. Figure (3-3) shows the volume of global demand for the most demanded polymers in 2005. It is clear that polyethylene is certainly the most important among these polymers. The total quantity of the three types of polyethylene represents more than 39% of the polymers consumed in the World. This is a towering figure that exceeds several billion tons annually.

Manufacturing polyethylene necessitates the use of ethylene at an ultra high purity (99.95% or more).

There are three main types of polyethylene, LDPE, LLDPE and HDPE:

a) Low-density polyethylene, LDPE,
b) Linear low-density ethylene, LLDPE,
c) High-density polyethylene, HDPE.

The first has been discovered by accident in 1933 by Faust and Jason from the British Imperial Company, ICI, during a study of the reaction between ethylene and benzaldehyde, under high pressure. It is known as low-density polyethylene or high-pressure polyethylene. LDPE is manufactured by heating pure ethylene at 200°C, and under more than 1500 atmospheric pressures, in the presence of a polymerization catalyst of 0.03% to 0.01% of oxygen or peroxide to give a yield of 20 and 30%, respectively:

$$n\ CH_2=CH_2 \longrightarrow \left[CH_2\text{-}CH_2 \right]_n$$

The molecules of polyethylene manufactured by this method are characterized by the presence of many branches (a branch in almost every 50 carbons). These branches prohibit the polymer molecules

47

from stacking in parallel with each other, hence the low density. Most branches are formed by the reaction of free radicals at the terminal of the polymer by abstracting a hydrogen atom from the long chain giving a small branch, three or four carbons long. The reaction starts another new free radical in the center of the molecule as follows:

$$
\begin{array}{ccc}
& H_2C\text{---}CH_2 & \\
& / \quad\quad \backslash & \\
\sim\text{---}CH_2\text{---}CH & \quad CH_2 & \longrightarrow \\
& \backslash \quad\quad \cdot/ & \\
& H\text{----}CH_2 &
\end{array}
\qquad
\begin{array}{cc}
& H_2C\text{---}CH_2 \\
& / \quad\quad \backslash \\
\sim\text{---}CH_2\text{---}\overset{\cdot}{C}H & \quad CH_2 \\
& / \\
& H_3C
\end{array}
$$

$$
\xrightarrow{\;5\,CH_2=CH_2\;}
\quad
\begin{array}{c}
CH_2CH_2CH_2CH_3 \\
/ \\
\sim\text{---}CH_2\text{---}CH(CH_2CH_2)_4CH_2\overset{\cdot}{C}H_2
\end{array}
$$

In the manufacture of high-density polyethylene, the reaction is carried out in the presence of a coordination catalyst, such as "Ziegler-Natta" catalysts which consists of triethyl aluminum, $Al(Et)_3$ and titanium tetrachloride, $TiCl_4$, under a mild pressure of 6-7 atmospheres at a temperature of 60 to 70°C, and in the presence of a hydrocarbon solvent such as diesel oil, heptane, or toluene. Because oxygen and water affect the activity of the catalyst; the reaction is performed in an inert atmosphere such as nitrogen gas:

$$
nCH_2=CH_2 \longrightarrow \quad \left[CH_2\text{-}CH_2 \right]_n
$$

Linear compounds are obtained in these circumstances that stack easily to give high-density polyethylene. Because the compounds are straight in these polymers they are more crystalline, elaborate, stronger, solider, and more tolerant to extrusion than the branched low-density polyethylene.

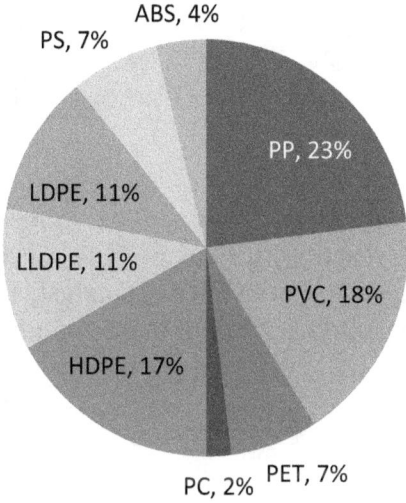

Figure 3-3: The volume of the global demand for polymers, 2005.

It is agreed that polymerization in the case of "Ziegler-Natta" catalyst takes place at specific effective centers on the surface of the catalyst. One of the most widely accepted mechanisms of this reaction is the single metal mechanism:

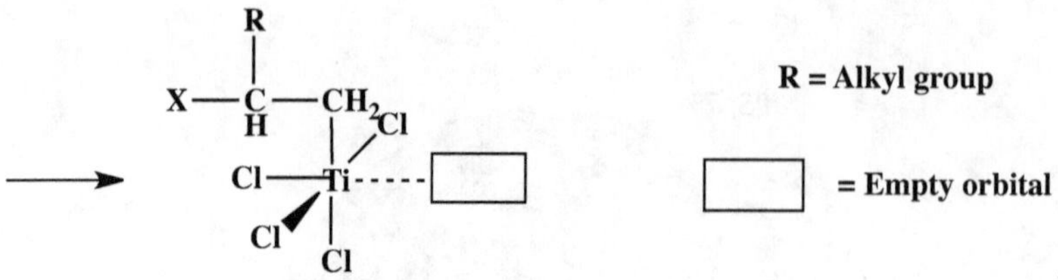

R = Alkyl group

☐ = Empty orbital

Although the single-metal mechanism gives a reasonable interpretation of the reaction, it does not explain the importance of presence of the aluminum atom in the catalyst, and hence, a two-metal reaction mechanism has been proposed as illustrated below:

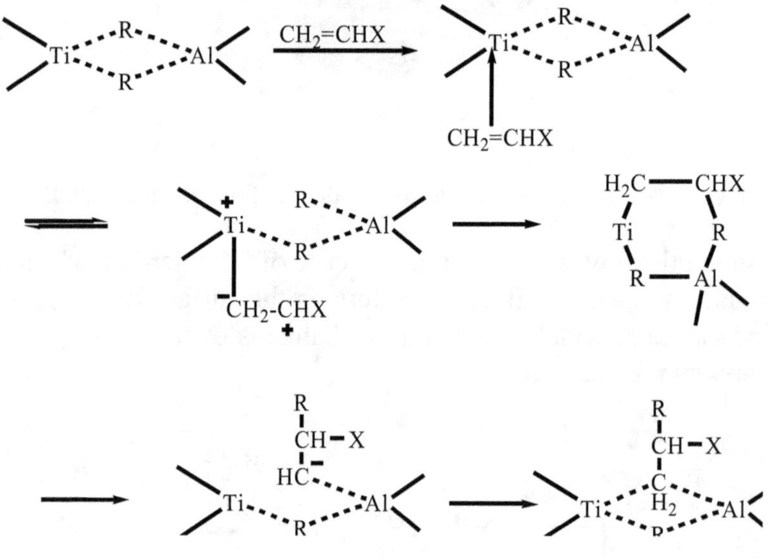

Table (3-6) summarizes the most prominent natural properties of the polyethylenes: high-density, low-density, and linear low-density polyethylene. Polyethylenes are the most valuable commercial plastics. They have multiple uses in the chemical industry, household and consumer goods, electrical insulation, packaging reels, toys, pipes and others.

50

Table 3-6: The most prominent natural properties of polyethylenes.

Property	LDPE	LLDPE	HDPE
Density gm./cm^3	0.92	0.94	0.96
Crystalline m. p., °C	108	123	133
Tensile strength lb./in^2	1500	1600	4000
Elongation at break (%)	450	700	500
Hardness (Shore D)	45	55	65
Softening point, Vicat C	95		120

Metallocenes

The global companies have developed many Ziegler-Natta catalysts to reach out to the second generation of catalysts known as metallocenes. The structure (Figure 3-4) illustrates one of the most notable metallocenes called zirconocene because of the existence of a zirconium atom between the two groups of cyclopentadiene. It can be noted that the chemical composition of the catalyst allows the monomer to enter from one direction only, thus giving polymer products of high quality, stable properties, and stereostructure. Metallocene catalysts necessitate the presence of methyl aluminum oxide (MAO) as a co-catalyst. Although the metallocenes are extremely expensive (five dollars per gram), each gram of them is sufficient to polymerize at least one metric ton of ethylene to polyethylene (Figure 3-4).

Table (3-7) summarizes a comparative study of the most significant new technologies used in the production of polyethylene, including the owner, product, density range, and the melt flow rate.

Table (3-8) summarizes the most beneficial effects of the second-generation of metallocenes on the properties of the produced polyethylenes.

51

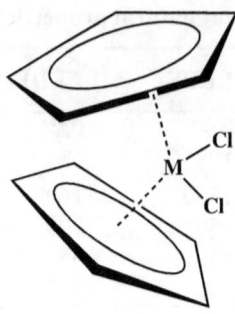

M = Zr, Ti, Hf....

Cocatalyst = Methyl aluminoxane)MAO from $Me_3Al + H_2O$

Figure 3-4: Metallocenes are the second generation of Ziegler-Natta catalysts.

The world is witnessing today an increasing demand for linear low-density polyethylene, LLDPE, compared to the low-density polyethylene, LDPE. Perhaps some of the reasons for this are the decline in operating and capital cost, lots of savings in energy, and in the spaces required, in addition to the flexibility required, in the production, quality of products, improved performance and ease of transfer for the production of HDPE for example, with low losses of the reaction. For the abovementioned reasons, the demand for LLDPE increased.

LLDPE is manufactured through polymerization of ethylene in the presence of a small percentage of small olefins such as 1-butene, 1-hexene or 1-octene over a metallocene catalyst.

$$n\ CH_2{=}CH_2 + 1/_{20}\ n\ CH_2{=}CHR \longrightarrow \left[CH_2CH_2(CH_2CHR)_{1/_{20}} \right]_n$$

Table 3-7: A comparative study of the most significant new technologies used in the production of polyethylene.

Owner	Product	Density gm/ml	Melt flow rate gm/10"
Basell Tech	ULDPE, VLDPE, LLDPE, HDPE	<0.900->0.960	0 .01-100
Borealis A/S	LLDPE, MDPE, HDPE	0.918-0.970	0 .1->100
BP	LLDPE, HDPE	-	-
Mitsui	MDPE, HDPE	0.930-0.970	0 .01->50
Univation	LLDPE, HDPE	0.915-0.970	0 .1->200

Table 3-8: The most serious effects of the second-generation metallocenes on properties of polyethylenes.

Catalyst	Main characteristics	Aim
Mono-site metallocene	Clear, transparent, easy processing	Wide use, low price, increase production
Di- and tri-site	Easy processing, transparent	Commodity
Wide range M. W.	Easy processing, developable	Wide utility
Superhexene	Improved mechanical properties	Using octane instead of hexane to produce LLDPE

Ethyl benzene

Ethyl benzene is made in the modern industry from benzene and pure ethylene or a mixture of ethane and ethylene in the presence of zeolite as a catalyst as follows:

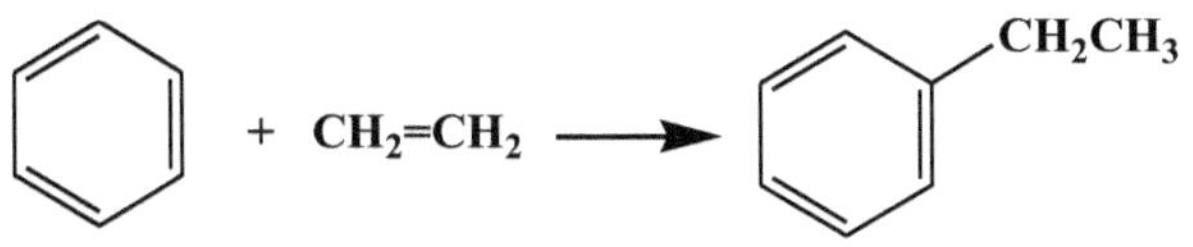

Benzene, in the liquid state, reacts with ethylene, in the alkylation reactor which contains several fixed layers of Mobil MCM-22 catalyst, to give ethyl benzene, and small quantities of polyethylbenzene (PEB) that are produced through the fast reaction of ethylene with ethyl benzene, because the ethyl group activates the benzene ring:

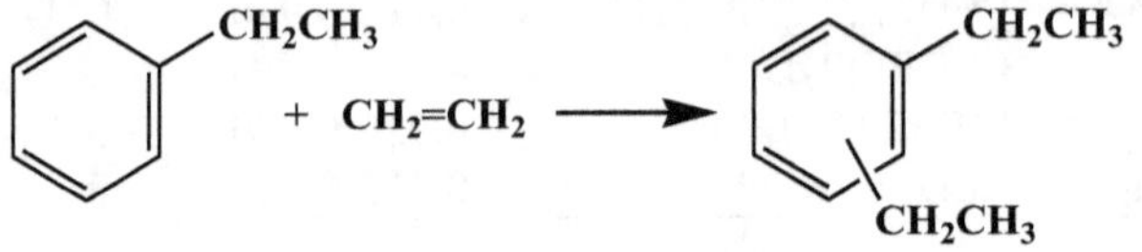

In the trans-alkylation reactor, polyalkylbenzenes are mutually converted into ethyl benzene over Mobil TRANS-4 catalyst:

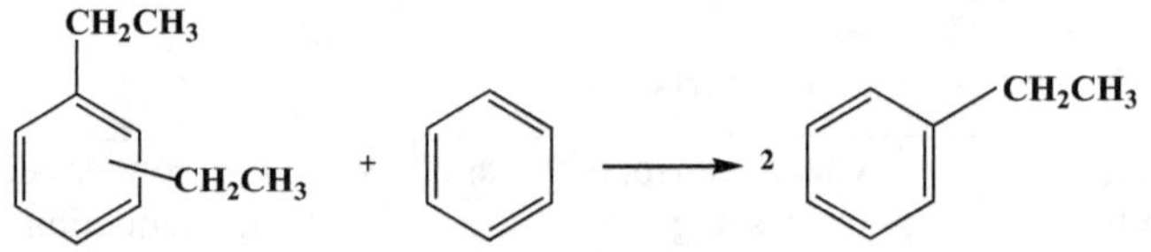

The outputs of the reactors are pumped into the benzene column where unreacted benzene is separated from crude ethyl benzene. The fresh benzene is pumped together with a part of the acquired benzene to the gas column to get rid of the light impurities. The heavy end obtained from the gas column is fed to the ethyl benzene column to extract ethyl benzene. From the bottom of this column polyethylbenzene (PEB) is withdrawn for recycling to the other column. Diethyl benzene is separated and used as fuel. This process produces highly pure ethyl benzene. The purity ranges between 99.95 and 99.99%.

Styrene

Styrene monomer is synthesized by the reaction of benzene and ethylene and then stripping the hydrogen from ethyl benzene by heat. The following equations explain the progress of the reaction:

The synthesis of ethyl benzene proceeds by the reaction of benzene with ethylene, in the liquid state, at 140-200°C, and under a pressure of 4-9 atmospheres. The ratio of 3:1 gives a high yield of ethyl benzene, and reduces the formation of polyethylbenzene. The reaction comes to an end in about half an hour in the presence of anhydrous aluminum chloride as a catalyst.

The synthesis of ethyl benzene in the gaseous state is not common, however, can be done in the presence of phosphoric acid, boron trifluoride or silica-alumina as a catalyst, at 300°C, and under 40-60 atmospheres of pressure, and the ratio of ethylene to benzene is 1:5.

The second step for obtaining styrene is the abstraction of hydrogen from ethyl benzene, which is performed at 600-660°C, and low pressure is obtained by the addition of superheated steam, in the presence of iron oxide as a catalyst, activated with chromium oxide and potassium carbonate. The styrene formed is separated from ethyl benzene, by vacuum distillation however, the addition of an inhibitor such as sulfur or another modern one, is necessary to prevent the polymerization of styrene during separation. p-Tri-butyl catechol is added as an inhibitor to the final product to prevent styrene from polymerization.

Since the first reaction is exothermic, the produced heat is used in transforming water into steam, which is used in the second endothermic reaction. It is natural to produce polyethylbenzene in the first reaction because ethyl benzene is more active than benzene;

therefore, polyethylbenzene is turned into ethyl benzene again by reaction with benzene through trans-alkylation as previously mentioned.

It is known that ethyl benzene can be obtained from the output of catalytic reforming and conversion of LPG into aromatics, and there is some ethyl benzene present in the pyrolysis gasoline too. Ethyl benzene can be deduced from these resources and used to produce more styrene monomer, among other products.

There are other methods of synthesis of styrene from ethyl benzene, including the abandoned traditional method, which starts with oxidation of ethyl benzene with air to acetophenone and ethyl phenyl alcohol. The hydrogenation of acetophenone gives ethyl phenyl alcohol, which can be dehydrated easily to produce styrene as illustrated in the following equations:

$$CH_2CH_3\text{-}C_6H_5 \xrightarrow{[O]} C_6H_5\text{-}CCH_3 (=O) \xrightarrow{[H]} HO\text{-}CHCH_3\text{-}C_6H_5 \xrightarrow{-H_2O} CH=CH_2\text{-}C_6H_5$$

The relatively modern method of synthesis of styrene from ethyl benzene is the method of Halcon. The styrene manufacturers have started using this method in the nineteen sixties. The method is characterized by the production of ethyl phenyl alcohol with the simultaneous conversion of propylene to propylene oxide. Ethyl phenyl alcohol is dehydrated to give styrene. This method is famous for using low temperatures compared to the other methods, which means saving energy, as well as producing smaller quantities of byproducts. In this method, ethyl benzene is oxidized by air, at 70°C, to give the ethyl benzene hydroperoxide, which reacts with propylene at 120°C, under 35 atmospheres, in the presence of dissolved molybdenum as a catalyst:

56

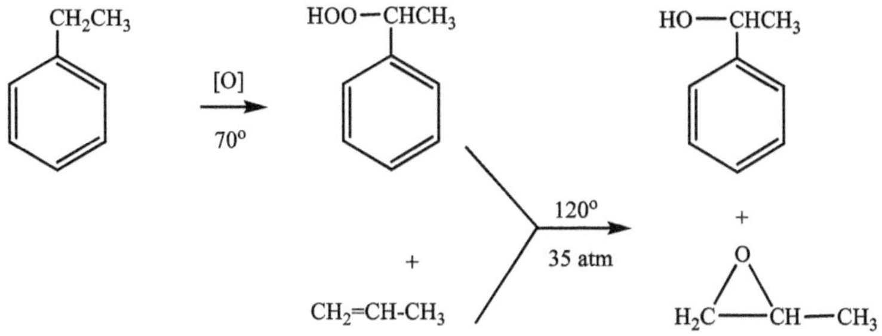

It seems that the mechanism of the reaction depends on the formation of a polarized complex between the catalyst and hydroperoxide. The electron deficient oxygen atom in the complex attacks the double bond in propylene, as illustrated in the following equation:

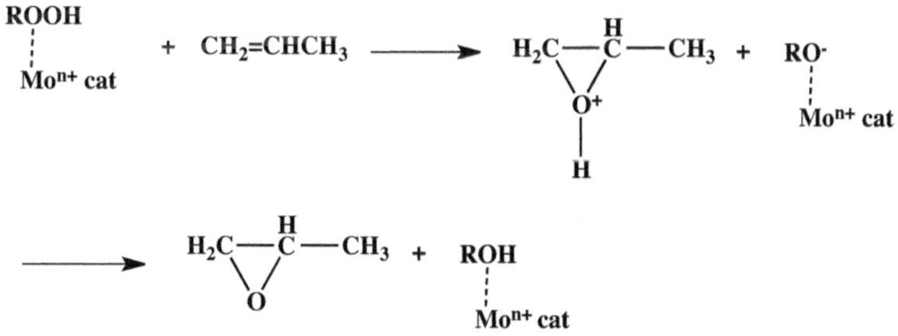

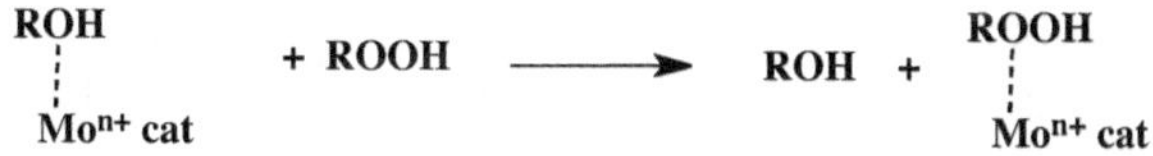

Ethyl phenyl alcohol can be dehydrated, in the presence of titanium dioxide as a catalyst, at 180-280°C to yield styrene:

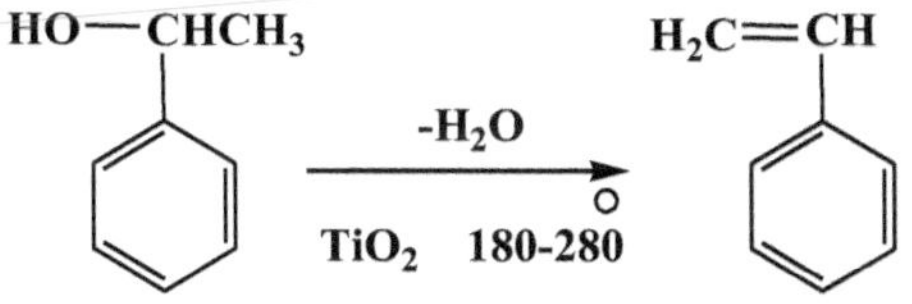

The world today consumes millions of tons of styrene annually used in the manufacture of polystyrene, acrylonitrile-butadiene-styrene (ABS) resins, styrene-acrylonitrile (SAN) resins, the synthetic styrene-butadiene rubber (SBR), and other copolymers with distinctive characteristics.

Polystyrene

Styrene is polymerized to polystyrene in the presence of benzoyl peroxide catalyst at 80-85°C:

$$n \begin{array}{c} CH=CH_2 \\ \bigcirc \end{array} \xrightarrow{C_6H_5COOCC_6H_5} \begin{array}{c} -[CHCH_2]_n \\ \bigcirc \end{array}$$

Polystyrene is one of the most valuable commercial plastics with multiple uses in different devices, household appliances, food containers, packaging, toys, thermal insulation, decoration and accessories. Styrofoam is used in shipping goods, glass wool reinforced resins and many other uses.

UOP has developed in collaboration with Nippon Steel Chemical Company (NSCC) a process for producing multi-grade polystyrene. It has already produced more than fifty degrees of General Purpose polystyrene for public use characterized by transparency, amorphous, used in food containers, compact disc manufacture, and so many different uses. In addition to the General Purpose polystyrene; this technology can produce Impact Resistant polystyrene required for uses such as computer screens and television homes. This type can be integrated with polybutadiene rubber matrix, to produce another family of excellent, high-strength, high-luster, and high-gloss polystyrene. This kind can replace the expensive poly(acrylonitrile-butadiene-styrene) (ABS) in many applications.

The polymerization process depends on the continuous process in the bulk-phase of styrene using a mixture of chemical catalysis and thermal stimulation. The basic design of the process includes an independent unit of General Purpose Polystyrene (GPPS), and another for High Collision High Impact Polystyrene (HIPS), each of which can be produced in multiple grades. The multi-grade general-purpose polystyrene ranges between low molecular weight high-flow, and high-molecular-weight heat-resistant grades. The high impact polystyrene, besides being multi-grade, is characterized by high grades, in addition to the ability of distribution of rubber on the matrix of polystyrene, in a superbly unprecedented way. It has the ability of complete control of the volume of rubber granules to less than one micron in diameter. The small rubber produced extremely strong high impact polystyrene characterized by soft texture and glossy surface competing poly(acrylonitrile-butadiene-styrene) (ABS) in the mechanical properties, touch of the surface smoothness and gloss.

UOP has established several factories for Nippon Steel Chemical Company of Japan in the capacity of fifty thousand metric tons per year of "General Purpose" polystyrene, and an equivalent amount High Impact polystyrene, each of them in varying grades.

Linear alkyl benzene

The detergent industry requires a long tail of linear alkyl that ranges between ten and fourteen carbons long on a benzene ring. Linear alkyl benzene is prepared by the alkylation of benzene with linear paraffins through an olefin over a solid heterogeneous catalyst as follows:

The details of the reaction are shown in the following steps:

1) Hydrogen is removed from the alkyl (feed) in Pacol reactor to give the linear olefin.

2) The effluents are separated from the reactor into a gas and liquid fractions in the separator.

3) Di-olefins are hydrogenated in the liquid fraction selectively to mono-olefin in the DeFine reactor.

4) The light ends are separated in the stripper.

5) A mixture of paraffin and mono-olefin are reacted with benzene in the fixed bed Detal reactor.

6) The product goes from the reactor to the separation section and the unreacted benzene is recycled to the reactor.

7) The unreacted paraffin is separated and recycled to the Pacol reactor.

8) Linear alkyl benzene is separated in the column from the heavy alkyls.

Some features such as non-pollution and production of no waste characterize this process. Catalysts used in this process are non-corrosive and does not need any specific precautions.

The linear alkyl benzene produced in this process has a bromine index of less than 10, which is subject to 99% sulfonation to produce the detergents. More than twenty-five factories in the world are using this UOP owned technology.

Linear alpha olefins

A number of leading chemical industries such as linear low-density polyethylene industry, LLDPE, poly alpha olefins and alcoholic plasticizers demand appropriate high purity alpha olefins.

The process depends on the oligomerization of high purity ethylene to get the required compounds, as follows:

$$CH_2=CH_2 \longrightarrow CH_2=CHCH_2CH_3$$

$$+ CH_2=CH(CH_2)_3CH_3$$

$$+ CH_2=CH(CH_2)_5CH_3$$

$$+ CH_2=CH(CH_2)_7CH_3$$

The reaction is performed in the liquid state over a high efficiency and selectivity heterogeneous catalyst, in the presence of an appropriate solvent. The spent catalyst is separated, and the liquid is distilled to recycle the unreacted ethylene to the reactor. The products are separated in a distillation tower to the high purity alpha olefins.

This process is characterized by simplicity, easy reaction conditions of temperature and pressure, and the simple steel devices used. The catalyst is not poisonous, and the reaction is smooth in its presence.

SABLIN new linear alpha olefins

Table (3-9) summarizes the distribution of alpha olefin products (weight %). These products can be set and adjusted, according to the market requirements, provided that a remarkably small quantity of polyethylene, with a high boiling point is produced as a byproduct.

61

The French Petroleum Institute (IFP) owns this technology. IFP is famous for its extraordinary skills in heterogeneous catalysis.

Saudi Arabia Basic Industries Corporation (SABIC) has developed, in collaboration with the German Linde, a method for the synthesis of linear alpha olefins called SABLIN (Figure 3-5). Ethylene is used as a feedstock in a single step for the production of linear alpha olefins C_4-C_{20}, according to the following equation:

$$CH_2=CH_2 \longrightarrow CH_2=CH(CH_2)_nCH_3$$

$$n = 1\text{-}17+$$

The highly selective one step reactor gives high-purity products so that no more purification such as super-fractionation distillation is needed to remove any side products. Linear alpha olefins are subjected to a series of well-known traditional columns, to obtain the required final products.

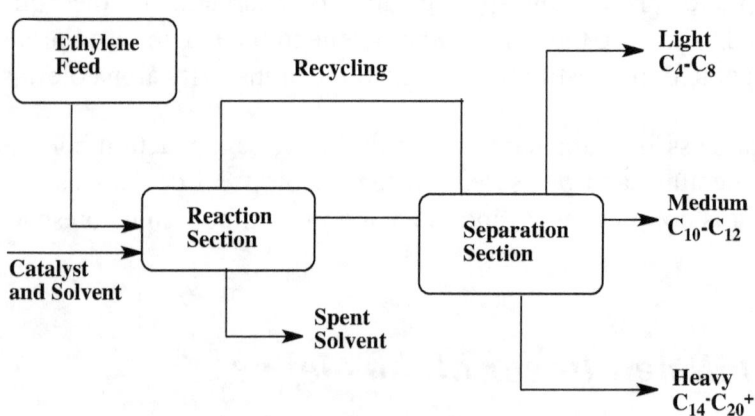

Figure 3-5: SABLIN new synthesis of alpha olefins.

Table 3-9: The distribution of alpha olefin products (weight %).

Product	Ratio, Wt. %
n-1-butene	33-43
n-1-hexene	30-32
n-1-octene	17-21
n-1-decene	9-14

The technical characteristics of this process include:

1) Low oligomerization heat and pressure.

2) The possibility of improved control over the distribution of the molecular weight of the reaction products.

3) High selectivity and purity of the products because of the use of non-noble metal catalysts.

The production costs of linear alpha olefins are low, as a result of lower operation conditions, low byproducts production, and lower consumption of utilities of water and electricity, together with low capital cost. This technology has environmental benefits as well because of the low waste products, and the possibility of recycling the undesirable products.

SABIC and Linde have established the first plant based on this technology in Jubail, Saudi Arabia, for the benefit of United Saudi Petrochemical Company with a production capacity of one hundred and fifty thousand tons yearly. Operation began in the first quarter in 2004.

The linear alpha olefins different applications include polymers, detergents, lubricating oils, fertilizers, and others. The light fractions such as 1-butene, 1-hexane, and 1-octene are crucial as co-monomers in the production of linear low-density polyethylene (LLDPE). The

World demand would double for these products due to increased worldwide demand for LLDPE.

The medium length alpha olefins fractions such as 1-undecene, 1-dodecene and 1-tetradecene are used to produce synthetic oils additives and detergents. The heavy linear alpha olefins fraction, such as 1-hexadecene, 1-octadecene and 1-ecocene arise in the production of lubricating oil additives, oilfield chemicals and wax products applications.

Ethanol (Ethyl alcohol)

More than 95% of the ethyl alcohol needed for the chemical industry in the modern era is synthesized using pure ethylene in two ways: first place in the liquid state at 30°C under the pressure of 20-30 atmospheres, where ethylene reacts with conc. sulfuric acid and then diluted with water as follows:

$$CH_2=CH_2 + H_2SO_4 \longrightarrow CH_3CH_2OSO_2OH$$

$$2CH_2=CH_2 + H_2SO_4 \longrightarrow CH_3CH_2OSO_2OCH_2CH_3$$

$$CH_3CH_2OSO_2OH + CH_3CH_2OSO_2OCH_2CH_3 + 3H_2O$$

$$\longrightarrow 3CH_3CH_2OH + 2H_2SO_4$$

The byproduct obtained in this reaction is ethyl ether as follows:

$$2CH_3CH_2OH + H_2SO_4 \longrightarrow CH_3CH_2OCH_2CH_3 + H_2SO_4.H_2O$$

Ethyl alcohol produced by this method is 95.6% pure with 4.4% water. For absolute ethyl alcohol, the product must be subjected to re-distillation with benzene.

The other method is the direct hydration of ethylene in the gaseous state, under the pressure of 68 atmospheres and 300°C, over phosphoric acid as a catalyst, on inert Calceleight or silica gel:

$$CH_2=CH_2 + H_2O \longrightarrow CH_3CH_2OH$$

There is no doubt that the traditional method of fermentation for the synthesis of ethyl alcohol is still alive. The ethyl alcohol produced was used in the synthesis of other chemicals such as ethylene in Pakistan, India, Peru, Brazil and many other developing countries. Also, it is used in the synthesis of butadiene, acetaldehyde, ethyl chloride, ethyl acetate, and others. Although fermentation depends upon cheap and available raw materials, the synthesis of some chemicals from ethyl alcohol is not economical, but the availability of alcohol locally, without the need to import it, offsets the economic difference.

Table 3-10: The conditions of reaction of olefins with sulfuric acid, for the synthesis of alcohols.

Olefin	Acid Concentration %	Temp. Range °C	Product
Ethylene	90-98	Up to 85	CH_3CH_2OH
Propylene	70-85	15-25	$CH_3CH(OH)CH_3$
n-Butene	70-85	20-30	$CH_3CH(OH)CH_2CH_3$
i-Butene	50-65	15-25	$(CH_3)_3COH$

Table (3-10) summarizes the conditions of reaction of olefins with sulfuric acid, for the synthesis of ethanol, propanol, isobutanol and tertiary butanol.

With the increased demand for motor gasoline in particular, the terrible contamination of the environment that lead to climate changes taking place in the globe, and the soaring prices of crude oil, the scientists started thinking about alternatives. One of the accepted ideas recently, is the addition of a proportion of ethanol to motor gasoline, the so-called green fuel, which meets the demands of the groups of activists seeking the prevention of environmental pollution. This trend began in Brazil, which acquires large quantities of ethanol from the fermentation of sugar cane residues, and is added to motor gasoline. The percentage of ethanol kept increasing in motor gasoline gradually until it reached the extent of running vehicles using ethanol alone. Although the fuel proved reasonably good, the performance of the engines reduced by half.

We have reached the green fuel and some countries began to subject the remnants of corn straw to fermentation for the synthesis of ethanol. The product required is in the global market, however it is using the animals feed. The same is true with regard to man and the poor suffered from the immoral idea that severely affected the lives of farmers and the food prices have doubled, as a result.

The idea of brewing the remnants of wood triggered the scientists to try upgrading the entire process, to include all the hydrocarbons available.

Acetaldehyde

The most critical operations in the acetaldehyde synthesis are the abstraction of hydrogen from ethanol by oxidation and the hydration of acetylene. The old method that seemed to diminish is the partial oxidation of ethylene over noble metal catalysts in line with the oxidation and reduction.

The truth is that the importance of acetaldehyde began to dwindle after some of its derivatives were obtained from other sources as side products, such as availability of acetic acid from methanol using Monsanto process, which will be discussed elsewhere in this book.

The process of obtaining acetaldehyde by the removal of hydrogen from ethanol is performed in two methods:

The first is by the removal of hydrogen in an endothermic process preferably over a catalyst of copper metal with the formation of hydrogen:

$$CH_3CH_2OH \xrightarrow{\text{cat.}} CH_3CHO + H_2$$

The second is by the removal of hydrogen in an exothermic process over a silver metal catalyst with the formation of water:

$$CH_3CH_2OH \xrightarrow{\text{cat.}} CH_3CHO + H_2O$$

The two processes are characterized by incompletion of the reaction; only 30-50% of ethanol is converted, therefore, separation, distillation and recycling processes are necessary. In economic terms, a balanced consumption of heat would be preferable by the addition of a calculated amount of air to evolve the required amount of heat to complete the reaction in a balanced manner.

As for the process of the preparation of acetaldehyde from acetylene, it has diminished because of the precious price of the acetylene, but it seems that the industry would return quickly to this method with the rapid rise of crude oil prices on the World market. The following equation summarizes the acetaldehyde synthesis from acetylene in the presence of sulfuric acid and magnesium sulfate as a catalyst:

$$HC\equiv CH + H_2O \longrightarrow CH_3CHO$$

The following equation shows the partial oxidation of ethylene in the presence of palladium chloride, which is highly selective:

$$CH_2=CH_2 + 0.5\ O_2 \xrightarrow{\text{Pd-Cu-cat.}} CH_3CHO$$

The oxidation mechanism in this process is characterized by the movement of OH group from the palladium to the ethylene carbon, namely so that all acetaldehyde oxygens come from the water medium:

$$\left[\begin{array}{c} Cl \\ \quad Pd \quad OH \\ Cl \quad CH_2 \\ H_2C \end{array}\right]^{\ominus} \longrightarrow \left[\begin{array}{c} Cl \\ \quad Pd-C_2H_4OH \\ Cl \end{array}\right]^{\ominus} \longrightarrow CH_3CHO + Pd^0 + 2\overset{\ominus}{Cl} + \overset{\oplus}{H}$$

The equation can be divided into two sections; the first represents the rapid oxidation of the olefin as follows:

$$CH_2=CH_2 + PdCl_2 + H_2O \longrightarrow CH_3CHO + Pd + 2\,HCl$$

The cupric chloride re-oxidizes the palladium metal to palladium chloride, which is restored to the second oxidation state of oxygen.

$$Pd + 2\,CuCl_2 \longrightarrow PdCl_2 + 2\,CuCl$$

$$2\,CuCl + 1/_2\,O_2 + 2\,HCl \longrightarrow 2\,CuCl_2 + H_2O$$

Figure (3-6) summarizes the most notable byproducts of acetaldehyde.

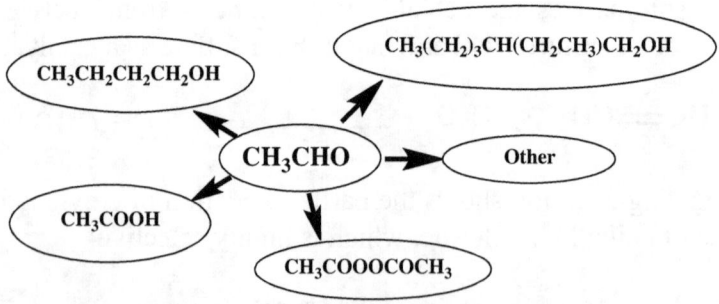

Figure 3-6: The most notable byproducts of acetaldehyde.

Acetic acid

Acetic acid is the first acid known to man in history. It is one of the most prominent aliphatic compounds demanded worldwide. The world consumption of acetic acid exceeds ten million tons per year.

Acetic acid is prepared in three ways:

a) From similar compounds with the same number of carbon atoms like acetaldehyde.
b) From larger compounds than its number of carbons by oxidation of n-butane or n-butenes.
c) Or by carbonylation of methanol. This last method is preferable recently and receiving a global boom due to the availability and cheapness of methanol.

Synthesis of acetic acid by oxidation of acetaldehyde

Acetic acid is prepared by air or oxygen oxidation of acetaldehyde through peroxyacetic acid through a free radical mechanism:

$$CH_3CHO + HOO\overset{\overset{\displaystyle O}{\|}}{C}CH_3 \longrightarrow 2CH_3COOH$$

The mechanism of formation of peroxyacetic acid proceeds as follows:

$$CH_3CHO + X^{\cdot} \xrightarrow[-HX]{} CH_3\overset{\cdot}{C}O \xrightarrow{+O_2}$$

$$CH_3\overset{\overset{\displaystyle \cdot}{\|}}{\underset{O}{C}}OO \xrightarrow[- CH_3\overset{\cdot}{C}O]{CH_3CHO} CH_3\overset{\overset{\displaystyle }{\|}}{\underset{O}{C}}OOH$$

The oxidation and reduction catalysts work in two steps; in the first the acetyl is formed through a free radical mechanism as follows:

$$CH_3CHO + M^{3+} \longrightarrow CH_3CO^• + M^{2+} + H^+$$

In the second step, the destruction of peroxyacetic acid proceeds as follows:

$$CH_3\overset{\overset{\displaystyle O}{\|}}{C}OOH + M_2^+ \longrightarrow CH_3\overset{\overset{\displaystyle O}{\|}}{C}O^• + M^{3+} + OH^-$$

The usual catalysts for these reactions are cobalt and molybdenum acetate.

In the case of using oxygen for oxidation rather than air, more cost will be incurred for its separation from the air, but it is favorable due to the high purity of the final products. When air is used, the cost becomes lower in one hand but then comes the need to remove acetaldehyde and acetic acid from the air coming out of the reaction. The two processes have the same selectivity and the same byproducts.

Synthesis of acetic acid by oxidation of alkanes and alkenes

The synthesis of acetic acid by the oxidative cracking of hydrocarbons is characterized by non-selectivity. The reaction mechanism proceeds via a free radicals system in a powerful exothermic process of C_4 to C_8 compounds and break them up to C_2, C_3 and C_4 compounds.

The large global companies who retained their respective patents have developed a number of ideas and applications. They are all characterized by non-selectivity to the extent that one of these processes produces acetic acid, in addition to eight other compounds.

70

The byproducts require multiple distillation operations, which are often corrosive and expensive.

Perhaps the best industrial preparation of acetic acid is from 1- or 2-n-butenes, whether cis or trans, known as Bayer method in two steps. The intermediate is acetic acid ester, which is catalyzed by protons for the addition of acetic acid to n-butane to give 2-acetoxybutane as follows:

$$CH_3CH_2CH{=}CH_2 + CH_3CH{=}CHCH_3 \xrightarrow{\text{AcOH/H}^+} 2\ CH_3CH_2\underset{\underset{CH_3}{|}}{CH}O\overset{\overset{O}{\|}}{C}CH_3$$

Acetic acid is added at 110°C, and 20 atmospheres, over an acidic ion exchange resin, with sulfonic acid groups. 1-Butene is isomerized to 2-butene, as a single product.

In the second step, the oxidation and cracking of 2-acetoxybutane to acetic acid by air takes place at 200°C and 60 atmosphere, without a catalyst:

$$CH_3CH_2\underset{\underset{CH_3}{|}}{CH}O\overset{\overset{O}{\|}}{C}CH_3 \longrightarrow 3\ CH_3CHO$$

Synthesis of acetic acid from methanol

This method is motivated by the availability of methanol in the world market in large quantities and for low prices besides the technical and economical viability of the process. This method has been implemented in the World. The future will witness a turnout valuable view due to the easy synthesis of methanol from methane, the main component of natural gas, over a cobalt iodide catalyst in the liquid state at 250°C and 670 atmospheres, as seen in the following equation:

$$CH_3OH + CO \xrightarrow{\text{cat.}} AcOH$$

The reaction mechanism is shown in the following steps:

$$2\,CoI_2 + 3H_2O + 11CO \longrightarrow 2\,H\text{-}Co(CO)_4 + 3\,CO_2 + 4\,HI$$

$$HI + CH_3OH \longrightarrow CH_3I + H_2O$$

$$H\text{-}Co(CO)_4 \xrightarrow{CH_3I} CH_3\text{-}Co(CO)_4 + HI$$

$$CH_3\text{-}Co(CO)_4 \xrightarrow{CO} CH_3CO\text{-}Co(CO)_4$$

$$CH_3CO\text{-}Co(CO)_4 \xrightarrow{H_2O} CH_3COOH + H\text{-}Co(CO)_4$$

This process is characterized by high selectivity, and a few side-products in small quantities. Carbon dioxide is produced in the process.

In a remarkable development of this method, Monsanto has replaced cobalt by rhenium, Rh, to give high selectivity, and the side products are carbon dioxide and hydrogen. The operation takes place at atmospheric pressure or slightly higher, and the reaction mechanism as in the case of cobalt iodide and CH_3RhCO becomes the active type as in the following equation:

$$[CH_3\text{-}Rh(CO)_2I_3]^- \xrightarrow{CO} [CH_3CO\text{-}Rh(CO)_2I_3]^-$$

$$\xrightarrow{CH_3OH} CH_3COOH + [CH_3\text{-}Rh(CO)_2I_3]^-$$

Uses of acetic acid

The most valuable uses of acetic acid are in the manufacture of various esters, which have numerous uses.

Acetic acid esters are used as monomers such as vinyl acetate and in the synthesis of rayon, which is cellulose acetate, manufactured from acetic acid. The acetate esters are used as solvents in the paint and resins industries, such as methyl, ethyl, isopropyl and butyl acetate.

Inorganic acetate salts are used in the manufacture of dyes, clothing and medicine.

Chloroacetic acid is used in synthetic organic chemistry. The raw materials needed for the synthesis of the anhydride are acetic anhydride and ketene.

Table (3-11) summarizes the most valuable uses of acetic acid in 2000 (Wt. %).

Table 3-11: The most valuable uses of acetic acid in 2000 (Wt. %).

Product	Ratio, wt %
Vinyl acetate	51
Cellulose acetate	5
Acetic acid esters	13
Acetanilide, acetic anhydride, acetyl chloride and acetamide	16
Solvent in terephthalic acid and dimethyl terephthalate	8
Chloroacetic acid	3
Inorganic acetate salts: Al, NH_4, Na, K etc.	4
Total	**100**

Ethyl acetate

Ethyl acetate is prepared by one of the following three methods. The selection is governed by the availability of raw materials, which vary from one country to another. Perhaps the easiest way is the esterification of acetic acid with ethanol in the presence of an acid as a catalyst:

$$CH_3CH_2OH + CH_3\overset{\overset{\displaystyle O}{\|}}{C}OH \xrightarrow{\;H^+\;} CH_3\overset{\overset{\displaystyle O}{\|}}{C}OCH_2CH_3 + H_2O$$

The synthesis of ethyl acetate from acetaldehyde through Chinko's method is adopted when acetaldehyde is abundant or the price of ethanol is too high. The reaction is conducted in the presence of ethyl aluminum as a catalyst and zinc ions and chlorine as activators:

$$2\ CH_3CHO \xrightarrow{\;Cat.\;} CH_3\overset{\overset{\displaystyle O}{\|}}{C}OCH_2CH_3$$

The third method is by the addition of acetic acid to ethylene according to the following equation:

$$CH_2{=}CH_2 + CH_3\overset{\overset{\displaystyle O}{\|}}{C}OH \longrightarrow CH_3\overset{\overset{\displaystyle O}{\|}}{C}OCH_2CH_3$$

Ethyl acetate is used in the manufacture of dyes as a solvent, in the chemical and pharmaceutical industries for extraction, such as extracting antibiotics.

Ethylene oxide

The importance of ethylene oxide originates from its use in the synthesis of ethylene glycol that is required in the manufacture of polyethylene terephthalate (PET). The synthesis of ethylene oxide starts from ethylene and oxygen, as in the following equation, in the presence of silver metal as a catalyst:

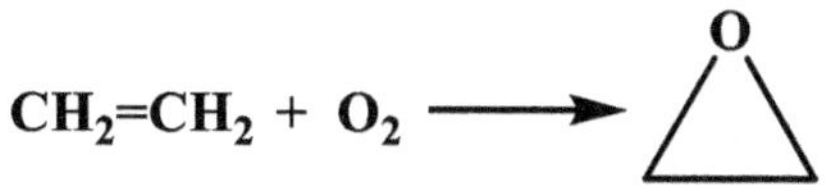

$$CH_2=CH_2 + O_2 \longrightarrow$$

Table 3-12: The fundamental differences between the three technologies of ethylene oxide.

Company	% of Int'l production	Yield %	Characteristics
Shell	43	140	Catalyst long life
Scientific Design	27	120	-
Union Carbide	20	-	Lower investment and low operation cost

Carbon dioxide is produced as a by-product from all the known technologies of preparation of ethylene oxide. Three companies are competing in the international market for ethylene oxide plants: Shell, Scientific Design, and Union Carbide. Table (3-12) summarizes the fundamental differences between the three technologies.

The traditional method of preparation of ethylene oxide has been through ethylene chlorohydrin, followed by abstraction of hydrochloric acid using a suitable cheap base, such as sodium hydroxide, or calcium hydroxide as illustrated by the following equations. This method suffers from extensive use of chlorine, which leads to the production of large quantities of common salt, sodium chloride, or calcium chloride, and a number of annoying by-products:

$$CH_2=CH_2 + Cl_2 + H_2O \longrightarrow \underset{\underset{OH\quad Cl}{|\quad\;\;|}}{H_2C-CH_2}$$

$$\underset{\underset{OH\quad Cl}{|\quad\;\;|}}{H_2C-CH_2} + NaOH \longrightarrow \overset{O}{\triangle} + NaCl + H_2O$$

The modern method of preparation of ethylene oxide is implemented through the reaction of ethylene with air or oxygen directly over a silver metal catalyst. In the case of using air, the temperature needed is 260-290°C, while only 230°C is needed when using oxygen. The best results were obtained by applying a pressure that ranges between 20 and 30 atmospheres, although the process can take place under normal atmospheric pressure:

$$CH_2=CH_2 + \tfrac{1}{2}O_2 \longrightarrow H_2C\overset{O}{\triangle}CH_2$$

Since ethylene can be oxidized into carbon dioxide and water, according to the equation below, appropriate action must be taken, and complete control is implemented, to minimize the occurrence of this undesirable reaction:

$$CH_2=CH_2 + 3O_2 \longrightarrow 2CO_2 + 2H_2O$$

Perhaps the most eminent derivative of ethylene oxide is ethylene glycol. The latter is used as a cooler in car radiators to prevent boiling in the summer, and prevent freezing in the winter. This is implemented by direct hydration of ethylene oxide in the liquid state under atmospheric pressure, and at 50-70°C, in the presence of a mineral acid as a catalyst, and in abundance of water, to curb the formation of polyglycols, such as diethylene glycol and triethylene glycol, which are of particular industrial importance and their preparation is discussed intensively elsewhere in this book:

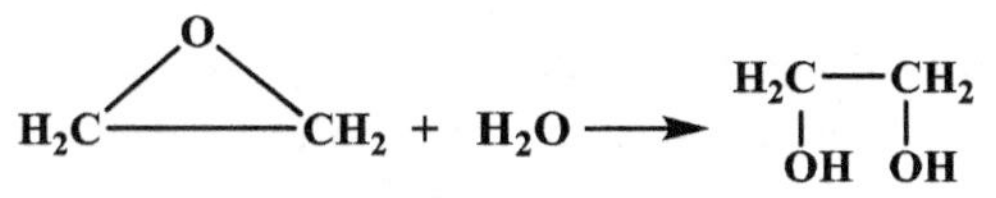

Ethylene glycol is of paramount importance in the synthesis of the famous polyethylene terephthalate polyester known commercially as Dacron:

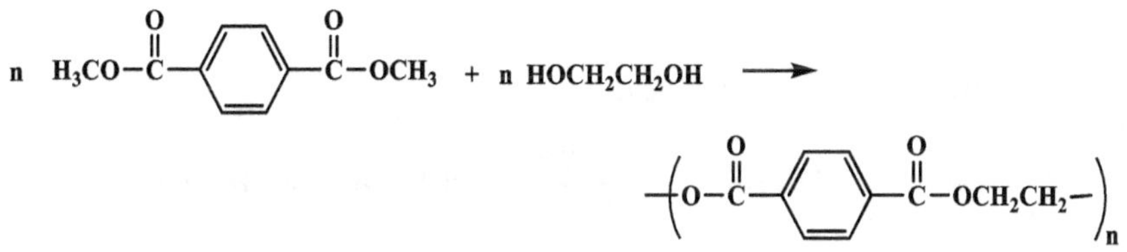

One of the most prominent industrial uses of ethylene oxide is its polymerization, in the presence of an acid or a base catalyst, for the synthesis of polyethers with alcoholic terminals that are used for the synthesis of polyurethane foams when reacted with diisocyanates such as toluene diisocyanate:

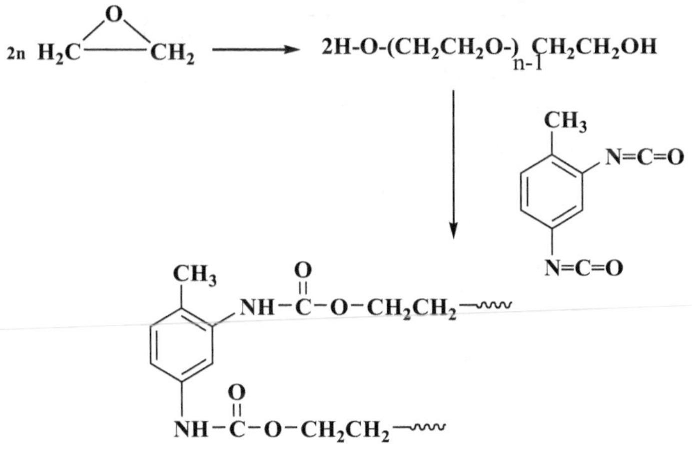

Polyurethane foam

The following methods are used in the past for the synthesis of acrylic acid, however they are abandoned now because of the inability of economic competition and the products were achieved from other various cheaper processes:

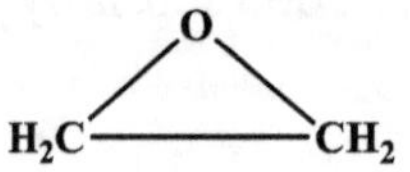

$$H_2C \overset{O}{\underset{\textstyle \diagup \diagdown}{\qquad}} CH_2 \quad + \; HCN \longrightarrow OHCH_2CH_2CN$$

$$OHCH_2CH_2CN \overset{-H_2O}{\longrightarrow} CH_2=CH-CN$$

$$CH_2=CH-CN + 2H_2O + H_2SO_4 \longrightarrow CH_2=CHCOOH + NH_4HSO_4$$

$$HOCH_2\text{·}CH_2CN + CH_3OH + H_2SO_4 \longrightarrow CH_2=CHCOOCH_3 + NH_4HSO_4$$

Table (3-13) summarizes the most notable uses of ethylene oxide in 2007 (ratio %).

Table 3-13: The most valuable uses of ethylene oxide in 2007 (ratio %).

Product	Consumption
Ethylene glycol	65
Di and triethylene glycol	7
Ethoxylates	13
Polyols	3
Ethanol amine	6
Ethylene glycol ethers	4
Polyethylene glycol	2
Total	100

Ethylene glycol

Ethylene oxide is used for the synthesis of ethylene glycol, diethylene glycol and triethylene glycol as in the following equations:

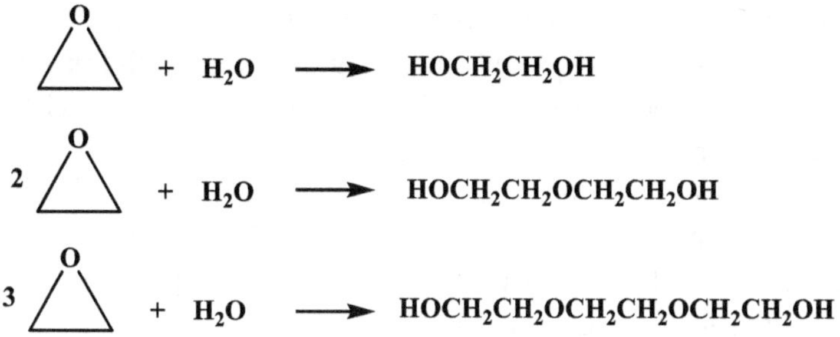

$$\triangle\text{O} + H_2O \longrightarrow HOCH_2CH_2OH$$

$$2\,\triangle\text{O} + H_2O \longrightarrow HOCH_2CH_2OCH_2CH_2OH$$

$$3\,\triangle\text{O} + H_2O \longrightarrow HOCH_2CH_2OCH_2CH_2OCH_2CH_2OH$$

The raw materials required to produce ethylene glycol are pure ethylene oxide and water. Ethylene glycol is used to prevent the freezing of liquids, and in polyester industry for the preparation of polyethylene terephthalate, PET, and the transparent plastic bottles manufactured from the aforementioned polyester. The diethylene and triethylene glycols are prepared according to the same procedure.

Ethylene chloride

For the preparation of ethylene chloride, ethylene is reacted with chlorine in the gas phase at 125°C, in the presence of iron chloride as a catalyst, which helps the formation of free chlorine radicals:

$$CH_2=CH_2 + Cl_2 \longrightarrow \begin{array}{cc} H_2C & CH_2 \\ | & | \\ Cl & Cl \end{array}$$

Ethylene chloride is used as a product and solvent at the same time in its preparation in the liquid state at 50°C.

In the relatively modern process of oxychlorination, air and hydrogen chloride are used as intermediates for the chlorination of ethylene into ethylene chloride:

$$CH_2=CH_2 + 1/_2\,O_2 + HCl \longrightarrow \begin{array}{cc} H_2C & CH_2 \\ | & | \\ Cl & Cl \end{array} + H_2O$$

This reaction is performed at atmospheric pressure, or a little higher, and a temperature ranging between 230 and 320°C, in the presence of cupric chloride and potassium chloride as a catalyst. The yield obtained is more than 90%. It is believed that the mechanism of reaction proceeds with the reaction of ethylene with cupric chloride to give ethylene chloride and then cupric chloride is regenerated when oxygen and hydrogen chloride react with cuprous chloride as outlined in the following equations:

$$CH_2=CH_2 + 2\,CuCl_2 \longrightarrow \underset{\underset{Cl}{|}}{H_2C}-\underset{\underset{Cl}{|}}{CH_2} + 2\,CuCl$$

$$Cu_2Cl_2 + 1/2\,O_2 \longrightarrow CuO.CuCl_2$$

$$CuO.\,CuCl_2 + 2HCl \longrightarrow 2CuCl_2 + H_2O$$

In an effort to develop the oxychlorination process, Kellogg Company has operated it in the liquid state at 180°C and a pressure of 18 atmospheres, in the presence of oxygen, hydrogen chloride, chlorine and ethylene. In this process, the oxychlorination and chlorination of ethylene take place simultaneously:

$$CH_2=CH_2 + Cl_2 \longrightarrow \underset{\underset{Cl}{|}}{CH_2}-\underset{\underset{Cl}{|}}{CH_2}$$

$$CH_2=CH_2 + 1/_2\,O_2 + 2\,HCl \longrightarrow \underset{\underset{Cl}{|}}{H_2C}-\underset{\underset{Cl}{|}}{CH_2} + H_2O$$

It is noted that the use of oxygen rather than air in the Kellogg process gave the greatest yield of products, and a longer lifetime for the catalyst, however this happened at the expense of an additional unit of oxygen separation from the air.

Ethanolamines

Ethanolamines (mono-, di- and tri-) are produced from ethylene oxide and ammonia as perceived in the following equations:

$$NH_3 \ + \ \triangle\!\!\!\!\!{}^{O} \longrightarrow H_2NCH_2CH_2OH$$

Monoethanolamine (MEA)

$$NH_3 \ + \ 2 \ \triangle\!\!\!\!\!{}^{O} \longrightarrow HN(CH_2CH_2OH)_2$$

Diethanolamine (DEA)

$$NH_3 \ + \ 3 \ \triangle\!\!\!\!\!{}^{O} \longrightarrow N(CH_2CH_2OH)_3$$

Triethanolamine (TEA)

The industrial process is explained in the following lines. Ammonia solution is fed with the recycled amine and ethylene oxide continuously at a controlled rate, to a reactor, which produces mono-, di- and triethanolamine. The proportion of products could be set to give the highest percentage of mono-ethanolamine. The reactor output contains, in addition to the three amines, water and ammonia, together with other amines as byproducts. In the separation section, ammonia is separated in a column and re-absorbed in the absorber and then recycled to the reactor. Water is separated in the evaporator and the dehydrator. Monoethanolamine is separated in the vacuum distillation column. Diethanolamine and triethanolamine are obtained similarly. The yield of this process amounts to more than 97% of the raw materials.

The owner of this technology is Acid-Amine Technologies Inc. It established several factories, producing four to twenty thousand tons per year.

Ethylene bromide

This chemical material is used in large quantities as additive to motor gasoline, containing tetraethyl lead, to raise the octane number. The addition of tetraethyl lead to motor gasoline is currently prohibited in the whole World because of its role in air pollution except for a few poor countries. It was noted that the failure to add ethylene bromide leads to the deposition of lead on the engine valves. Ethylene chloride is used as well in such use. Ethylene bromide is manufactured through the direct addition of bromine to ethylene:

$$CH_2{=}CH_2 \ + \ Br_2 \ \longrightarrow \ \underset{\underset{Br \ \ Br}{|} \ \ \ }{H_2C{-}CH_2}$$

While the general international trend is moving towards preventing the addition of tetraethyl lead to motor gasoline, and the production of unleaded motor gasoline, ethylene bromide has other useful derivatives, the most noteworthy of which is the synthesis of vinyl bromide that is used within the fiber and plastics industry as a flame retardant:

$$\underset{\underset{Br \ \ Br}{|}}{H_2C{-}CH_2} \ \xrightarrow{\text{-HBr}} \ \underset{\underset{Br}{|}}{CH_2{=}CH}$$

Vinyl chloride

Vinyl chloride monomer is synthesized recently from ethylene, chlorine and oxygen in two steps. The first step is the preparation of ethylene chloride from the reaction of chlorine with ethylene in the liquid state as follows:

$$CH_2{=}CH_2 \ + \ Cl_2 \ \xrightarrow{\text{cat.}} \ \underset{\underset{Cl \ \ Cl}{|}}{H_2C{-}CH_2} + \text{Heat}$$

Ethylene chloride is synthesized also through oxychlorination of ethylene by reaction with oxygen and hydrogen chloride as illustrated by the following equation:

$$CH_2=CH_2 + 2HCl + \tfrac{1}{2}O_2 \longrightarrow \underset{\underset{Cl}{|}}{H_2C}-\underset{\underset{Cl}{|}}{CH_2} + H_2O$$

Air can be used in oxychlorination of ethylene; however the use of oxygen gives better yields. Ethylene chloride is washed, dried, and filtered before use in the next step.

The second step in the preparation of vinyl chloride is the pyrolysis of ethylene chloride in a pyrolysis furnace to vinyl chloride and hydrogen chloride, as shown in the following equation:

$$\underset{\underset{Cl}{|}}{H_2C}-\underset{\underset{Cl}{|}}{CH_2} \longrightarrow CH_2=\underset{\underset{Cl}{|}}{CH} + HCl$$

Hydrogen chloride produced in this process is separated and pumped into a hydrochlorination reactor to be recycled for more ethylene chloride. The three reactions can be controlled so that only vinyl chloride is produced with continuous disposal and recycling of the hydrogen chloride produced.

This technology is applied in many factories, the total production of which well exceeds six million metric tons per year of vinyl chloride and more than fifteen million metric tons per year of ethylene chloride. The production capacity of these factories ranges between ten thousand and sixty thousand and more metric tons per year of vinyl chloride.

Vinyl chloride was synthesized in the early thirties of the twentieth century through the reaction of acetylene with hydrogen chloride in the presence of mercury chloride as a catalyst, at about 150°C, to give a yield that amounts to 90%:

$$CH\equiv CH \ + \ HCl \longrightarrow \underset{\underset{Cl}{|}}{CH_2}{=}CH$$

This process is condemned for using the expensive acetylene, which was prepared from calcium carbide. In the late thirties of the last century ethylene was used instead of acetylene for the synthesis of vinyl chloride. The chlorination of ethylene yields ethylene chloride which when followed by stripping the hydrogen chloride using a strong base such as aqueous sodium hydroxide or high temperatures yields vinyl chloride:

$$CH_2{=}CH_2 \ + \ Cl_2 \longrightarrow \underset{\underset{Cl \ \ \ Cl}{| \ \ \ |}}{H_2C{-}CH_2} \xrightarrow{500°C} \underset{\underset{Cl}{|}}{CH_2{=}CH} \ + \ HCl$$

Since the synthesis of vinyl chloride from acetylene needs hydrogen chloride, while the latter is artificially produced from ethylene chloride, some companies adopted using the two methods together. Although this solves the problem of the disposal of the noisy hydrogen chloride by-product, it increases the basic and operational costs, and uses the expensive acetylene.

Deacon method was used for the transformation of hydrogen chloride to chlorine, which reacts once more with ethylene for more ethylene chloride, and consequently vinyl chloride as shown in the following equation:

$$CH_2{=}CH_2 \ + \ Cl_2 \longrightarrow \underset{\underset{Cl \ \ \ Cl}{| \ \ \ |}}{H_2C{-}CH_2} \longrightarrow \underset{\underset{Cl}{|}}{CH_2{=}CH} \ + \ HCl$$

$$2HCl \ + \ 1/2O_2 \longrightarrow H_2O \ + \ Cl_2$$

Deacon reaction proceeds at 450°C, using copper chloride as a catalyst, and air as an oxidant.

The modern method to get rid of hydrogen chloride in the presence of ethylene is the oxychlorination method, which proceeds at about

300°C, in the presence of a catalyst of copper chloride and potassium chloride. The yield exceeds 90% as already mentioned.

One exotic method is discovered and developed by ABB Lummus Company, where ethane and chlorine are passed over copper oxychloride to give vinyl chloride in 80% yield based on ethane and 99% based on chlorine. The catalyst is reduced to cupric chloride, which is re-oxidized by air, to give copper oxychloride again.

Most of the vinyl chloride is polymerized into polyvinyl chloride (PVC) in the presence of a free radical catalyst. The most prominent method of polymerization is the suspension polymerization at about 50°C and 9 atmospheric pressures. Unfortunately, polymerization of vinyl chloride cannot be performed continuously, only batch processes are possible. Emulsion polymerization of vinyl chloride is used in certain applications where fine scale particles are required. Bulk or mass polymerization is used when pure and transparent products are needed. The suspension polymerization method of preparation is the most important one. Polyvinyl chloride is one of the most produced performance polymers. It has diverse and miscellaneous uses such as pipes, industry, tiles, cables, furniture, packaging and reels, compact discs, construction and transportation:

$$n \ CH_2=\underset{\underset{Cl}{|}}{CH} \longrightarrow -\!\!\left(CH_2-\underset{\underset{Cl}{|}}{\overset{\overset{H}{|}}{C}}\right)_n$$

Vinyl chloride is used also for the synthesis of vinylidene chloride in two steps: first, chlorination at 40°C, ending with 1,1,2-trichloroethane, and the second step is the abstraction of hydrogen chloride, using a cheap base, such as calcium hydroxide at 100°C, with high selectively of more than 90%:

$$CH_2=\underset{\underset{Cl}{|}}{CH} + Cl_2 \longrightarrow H_2\underset{\underset{Cl}{|}}{C}-\underset{\underset{Cl}{|}}{C}HCl \xrightarrow{Ca(OH)_2} CH_2=CCl_2$$

Vinylidene chloride is used for the synthesis of polyvinylidene chloride, known commercially as Saran, through copolymerization with vinyl chloride (15%) and 85% vinylidene chloride. Saran is produced in small quantities and used in the manufacture of packaging reels, and some fibers for the furniture industry through covering cellulose hydrate and polypropylene. Copolymerization of acrylonitrile 70% with 30% vinylidene chloride yields acrylic fibers that resist ignition.

Vinyl chloride is used in the synthesis of 1,1,1-trichloroethane when reacted with hydrogen chloride at 40°C, in the presence of ferric chloride as a catalyst followed by the reaction of the product with chlorine at 400°C, in a high yield approaching 95%.

$$CH_2=\underset{\underset{Cl}{|}}{C}H_2 \; + \; HCl \longrightarrow H_3C-\underset{\underset{Cl}{|}}{C}H_2 \xrightarrow[400\ C]{Cl_2} H_3C-CCl_3$$

1,1,1-Trichloroethane is synthesized from vinylidene chloride, which in turn is prepared from vinyl chloride at 30°C, in the presence of ferric chloride as a catalyst:

$$CH_2=CCl_2 \; + \; HCl \longrightarrow CH_3-CCl_3$$

Trichloroethane is used as a common industrial solvent in cleaning electrical and electronic devices. It competes with trichloroethylene in the removal of fat by vapor. It is less toxic, and perhaps this is one of the main reasons for its popular use.

1,1,2-Trichloroethylene

One method of the synthesis of trichloroethylene is through the reaction of chlorine with acetylene at 80°C, in the presence of a catalyst, such as ferric chloride and then, hydrogen chloride is abstracted thermally at 300-500°C, in the presence of activated carbon, or at 600°C, without a catalyst:

$$CH \equiv CH + Cl_2 \longrightarrow CHCl_2 - CHCl_2 \xrightarrow{-HCl} CHCl = CCl_2$$

The high price of acetylene directed attention recently to the synthesis of trichloroethylene from ethylene by chlorination to give 1,1,1,2-tetrachloroethane followed by the removal of hydrogen chloride thermally:

$$CH_2 = CH_2 \xrightarrow{Cl_2} CH_2Cl - CH_2Cl \xrightarrow[-HCl]{Cl_2} CH_2Cl - CCl_3 \xrightarrow{-HCl} CHCl = CCl_2$$

Since huge quantities of hydrogen chloride are produced in this reaction, oxychlorination is an appropriate substitute, where dichloroethane is reacted with chlorine and oxygen over a copper containing catalyst, at 425°C, to obtain a mixture of trichloroethylene and perchloroethylene:

$$\underset{\begin{matrix} | & | \\ Cl & Cl \end{matrix}}{CH_2\text{-}CH_2} + Cl_2 + O_2 \longrightarrow \underset{\begin{matrix} | \\ Cl \end{matrix}}{HC} = CCl_2 + Cl_2C = CCl_2 + H_2O$$

Perhaps the most important use of trichloroethylene is as a solvent, which is used to clean metals from grease and fats. It is used also in dry cleaning and anesthesia.

The disadvantage of trichloroethylene is that it produces corrosive and toxic materials when heated in an atmosphere of oxygen, and therefore some stabilizers are added to curb its disintegration, such as some amines and phenols:

$$2\ CHCl = CCl_2 \xrightarrow{O_2} Cl_2HC - \overset{\overset{\displaystyle O}{\|}}{C}Cl + HCl + CO + COCl_2$$

Perchloroethylene

Perchloroethylene was synthesized from acetylene by chlorination, followed by stripping hydrogen chloride, through the suspension of pentachloroethane in calcium hydroxide, or in the presence of a copper chloride catalyst, at 300°C:

$$CH\equiv CH \xrightarrow[-HCl]{Cl_2} CHCl=CCl_2 \xrightarrow{Cl_2} CHCl_2-CCl_3 \xrightarrow{-HCl} CCl_2=CCl_2$$

We have seen in the preceding paragraphs the synthesis of perchloroethylene through oxychlorination to dichloroethane rather than the precious acetylene. The chlorination of propane, or propene, in harsh conditions, such as 600°C, gave perchloroethylene in an economic way:

$$H_3C-CH_2-CH_3 \ + \ Cl_2 \xrightarrow{600^0} CCl_2=CCl_2 \ + \ CCl_4 \ + \ 8HCl$$

$$H_3C-CH=CH_2 \ + \ Cl_2 \xrightarrow{600^0} CCl_2=CCl_2 \ + \ CCl_4 \ + \ 6HCl$$

Carbon tetrachloride is the by-product obtained, however it can be turned over to perchloroethylene, in the same reaction conditions:

$$2\ CCl_4 \ \rightleftharpoons \ CCl_2=CCl_2 \ + \ 2Cl_2$$

Perchloroethylene is similar to trichloroethylene in properties, uses and thermal cracking products, in the presence of oxygen. It gives toxic and corrosive substances that could be discouraged by stabilizers:

$$2\ CCl_2=CCl_2 \ + \ O_2 \longrightarrow Cl_3C-\overset{\overset{\displaystyle O}{\|}}{C}Cl \ + \ 2\ COCl_2$$

Perchloroethylene is used mainly in dry cleaning with the advantage of not removing or affecting colors and as a solvent in numerous uses, but the high degree of boiling prevents its use in removing fat from metals, due to the need for cooling metals after washing.

Ethyl chloride

Ethyl chloride is synthesized by the addition of hydrogen chloride to ethylene in the presence of aluminum chloride as a catalyst in the gaseous state at 130-250°C, or in the liquid state at 40°C in a solvent such as ethylene chloride or ethyl chloride itself:

$$CH_2{=}CH_2 + HCl \longrightarrow CH_3CH_2Cl$$

Ethyl chloride is synthesized from ethane too by reaction with chlorine at 300-400°C in a yield approaching 75%:

$$CH_3{-}CH_3 + Cl_2 \longrightarrow CH_3CH_2Cl + HCl$$

Hydrogen chloride becomes a problem if not disposed of properly. In fact, the success of this process depends on the use and disposal of the hydrogen chloride produced.

In some countries, the so-called integrated method is used in terms of preparation of ethyl chloride using the two methods; thus mixing chlorine, ethane and ethylene at 400°C, so that chlorine reacts with ethane to give ethyl chloride, which is separated, and hydrogen chloride produced reacts with ethylene, for more ethyl chloride:

$$CH_2{=}CH_2 + CH_3{-}CH_3 + Cl_2 \xrightarrow{400\ °C} CH_2{=}CH_2 + CH_3{-}CH_2Cl + HCl$$

$$\xrightarrow{150\text{-}250\ °C} 2CH_3CH_2Cl$$

Ethyl chloride is synthesized from ethanol in some poor countries. It is the traditional method any case:

$$CH_3{-}CH_2OH + HCl \longrightarrow CH_3CH_2Cl + H_2O$$

Ethyl chloride is used in the production of tetraethyl lead, which is used in raising the octane number in motor gasoline, but this compound has been abandoned internationally because of its toxicity and environmental pollution:

$$4\,NaPb + CH_3CH_2Cl \longrightarrow Pb(CH_2CH_3)_4 + 4\,NaCl + 3\,Pb$$

$$2\,CH_3CH_2MgCl + 2\,CH_3CH_2Cl + Pb \xrightarrow{DC} Pb(CH_2CH_3)_4 + 2\,MgCl_2$$

Ethyl chloride is used in the manufacture of ethyl cellulose and as an alkylating agent in the entire chemical industry, as a solvent, extracting agent, and a disinfectant in the medical field.

Since tetraethyl lead causes contamination of the environment, the international trend tends to replace it with other materials that increase the octane number of motor gasoline, and improve its quality, without the pollution of the environment. Among these materials are methyl tertiary butyl ether, MTBE, and tricarbonyl dipentadiene magnesium complex.

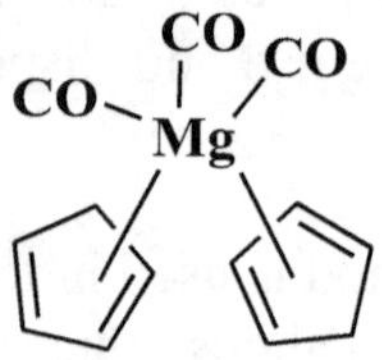

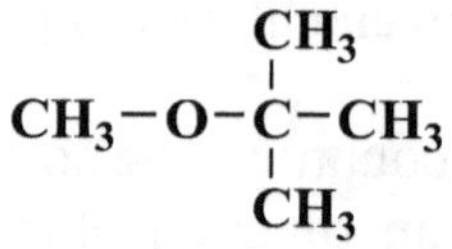

Vinyl fluoride

Vinyl fluoride can be synthesized from acetylene through the addition of hydrogen fluoride:

$$HC{\equiv}CH + HF \xrightarrow{cat.} CH_2{=}CHF$$

Vinyl fluoride can be synthesized through the catalytic replacement of chlorine by fluorine in vinyl chloride. 1-Chloro-1-flouroethane is formed as an intermediate and then hydrogen chloride is abstracted:

$$H_2C=CHCl \; + \; HF \; \xrightarrow{\text{cat.}} \; CH_3\text{-}CHClF \; \xrightarrow{\text{-HCl}} \; CH_2=CHF$$

Vinyl fluoride is polymerized to polyvinyl fluoride, which is used in painting in the form of a film or suspension. It resists harsh weather conditions.

Tetraflouroethylene

Tetraflouroethylene is a toxic gas that is polymerized to polytetraflouroethylene. It is the most significant fluorine plastic. Tetraflouroethylene is prepared by thermal stripping of hydrogen chloride from diflourodichloromethane with dimerization. It is believed that an intermediate of diflourocarbene is formed in the process:

$$2 \; CHClF_2 \; \longrightarrow \; F_2C=CF_2$$

To be polymerized, tetraflouroethylene must be very pure (99.9999%). The presence of a few parts per million of pollutants produces a polymer that is out of specifications and unfit for typical use.

Vinyl acetate

There are three methods of synthesis of vinyl acetate. They are summarized as follows:

1) From acetylene by reaction with acetic acid in a single and convenient selective step, but its use is limited to the countries that produce acetylene, which are the countries that have free hydropower:

$$HC\equiv CH + CH_3COOH \longrightarrow CH_2=CHO\overset{\overset{\displaystyle O}{||}}{C}CH_3$$

2) Vinyl acetate can be synthesized by reaction of acetic anhydride in two steps. In the first step ethylene acetate is formed as an intermediate, and in the second step acetic acid is abstracted in the presence of p-toluenesulfonic acid as a catalyst. The synthesis of vinyl acetate through this track is rarely used industrially:

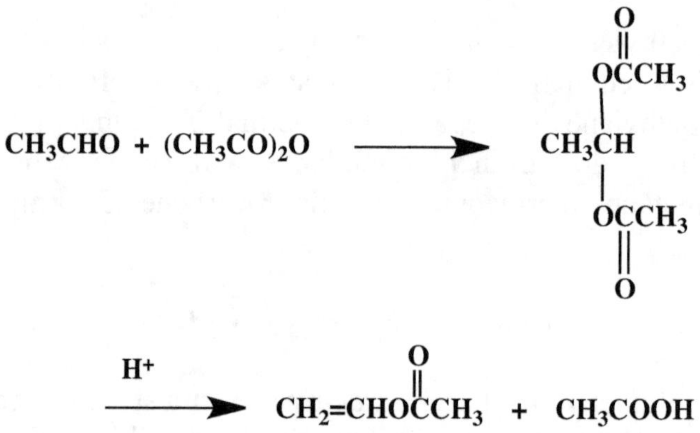

$$CH_3CHO + (CH_3CO)_2O \longrightarrow CH_3\overset{\overset{\displaystyle O\atop \displaystyle ||\atop \displaystyle OCCH_3}{|}}{\underset{\underset{\displaystyle O}{\overset{\displaystyle ||}{OCCH_3}}}{CH}}$$

$$\xrightarrow{H^+} CH_2=CHO\overset{\overset{\displaystyle O}{||}}{C}CH_3 + CH_3COOH$$

3) The synthesis of vinyl acetate from ethylene in one step is performed in the presence of palladium chloride as a catalyst:

$$CH_2=CH_2 + PdCl_2 + 2CH_3COONa \longrightarrow CH_2=CHO\overset{\overset{\displaystyle O}{||}}{C}CH_3 + NaCl + Pd + CH_3COOH$$

The mechanism of reaction proceeds as follows:

$$\begin{bmatrix} Cl & \\ & \diagdown \\ Cl\diagup & Pd \diagup^{OAc} \\ & \nwarrow_{CH_2} \\ & H_2C \end{bmatrix}^{\ominus} \longrightarrow \begin{bmatrix} Cl & \\ & \diagdown \\ Cl\diagup & Pd-C_2H_4OAc \end{bmatrix}^{\ominus} \longrightarrow CH_2{=}CHOAc + PdHCl_2$$

$$\longrightarrow \underset{H}{H_2C}{=}\underset{}{C}{-}O{-}\underset{\underset{O}{\parallel}}{C}{-}CH_3 \ + Pd^0 \ + 2\overset{\ominus}{Cl} \ + \overset{\oplus}{H}$$

In the presence of an oxidation and reduction system, palladium chloride can be reproduced from the palladium element formed, as seen in the oxidation of ethylene to acetaldehyde. The cupric ion will directly oxidize the palladium to the palladium ion. The cupric ion that was reduced to copper may be reproduced, in the presence of air, to the cupric ion. Thus, the outcome of the different reactions is summarized in the following equation:

$$CH_2{=}CH_2 \ + \ CH_3COOH \ + \ \tfrac{1}{2}O_2 \longrightarrow CH_2{=}CHO\overset{\overset{O}{\parallel}}{C}CH_3 \ + H_2O$$

More than half of the vinyl acetate in the World is used in the synthesis of polyvinyl acetate, which is used in the manufacture of raw materials for painting, and hydrated to polyvinyl alcohol, for the production of Vinylon fibers, stickers and paper glue. Table (3-14) summarizes the uses of polyvinyl acetate in the World in 2000 (Wt. %).

Table 3-14: The global uses of polyvinyl acetate in 2000 (Wt. %).

Product	Wt. %
Polyvinyl acetate and copolymers	52
Polyvinyl alcohol	30
Polyvinylchloride and copolymer vinyl acetate	4
Ethylene/vinyl acetate resins	4
Polyvinyl butyryl and other materials	10
Total	**100**

93

94

Chapter Four
Petrochemicals from Propylene

PROPYLENE 96

POLYPROPYLENE 100

ISOPROPYL ALCOHOL 101

ACETONE...102

METHYL ISOBUTYL KETONE 105

ACRYLIC ACID AND ITS ESTERS...106

METHACRYLIC ACID AND ITS ESTERS.......................................108

ACROLEIN 112

PROPYLENE OXIDE 113

ALLYL CHLORIDE 115

ALLYL ALCOHOL...116

GLYCERIN 118

ACRYLONITRILE 119

DI-, TRI- AND TETRAPROPYLENE..120

Propylene

Propylene is one of the most principal primary petrochemicals. The best significant uses of propylene are polymerization to polypropylene, and the production of acrylonitrile, to produce polyacrylonitrile. Table (4-1) summarizes the supreme essential uses of propylene in the World in 2000 (weight %) with total consumption exceeding fifty million tons. Table (4-2) summarizes the comparative demand of North America, Western Europe, and South East Asia for propylene in 2003 (weight %).

Perhaps the easiest approach for the synthesis of propylene is by the abstraction of hydrogen from propane as illustrated by the following equation:

$$CH_3CH_2CH_3 \longrightarrow CH_2=CHCH_3 + H_2$$

Table 4-1: The most important uses of propylene in the world in 2000 (Wt. %).

Product	Consumption
Polypropylene	56
Acrylonitrile	11
Propylene oxide	7
Isopropyl alcohol	3
Cumene	6
Oxo products	7
Oligomers	3
Other	7
Total	**100**

Table 4-2: A summary of the comparative demand of North America, Western Europe and South East Asia for propylene in 2003 (wt. %).

Product	N. America	W. Europe	S. E. Asia
Polypropylene	57	61	64
Acrylonitrile	11	7	15
Oxo alcohols	7	8	7
Propylene oxide	11	10	4
Cumene	1	6	4
Other	13	8	6
Total	**100**	**100**	**100**

Since this reaction is endothermic, it uses some energy and a continuously regenerated catalyst, according to UOP's famous technology that is used for the establishment of tens of factories worldwide, producing more than one million metric tons per year of propylene.

Another method for the production of propylene from C_4 fraction can be achieved using steam cracking. In this process, butadiene is hydrogenated to 1-butene, which is isomerized to 2-butene. Isobutene is separated for use in other preparations or treated with methanol to provide methyl tertiary butyl ether. The remaining portion of C_4, which is now 2-butene is treated with ethylene to produce propylene in the metathesis process, as shown in the following equation:

$$
\begin{array}{c}
H_3C \diagdown \quad \diagup CH_3 \\
HC = CH \\
+ \\
H_2C = CH_2
\end{array}
\quad \longrightarrow \quad 2\ CH_2 = CHCH_3
$$

The French Petroleum Institute, IFP owns this technology of metathesis. More than a hundred units were established around the World. This process would be more attractive when propylene is in high demand and butadiene is in less demand or produced in huge quantities.

ABB Lummus Company developed another technology for the production of propylene from ethylene and butene through a metathesis process. Butene can be obtained from dimerization of ethylene, or from other sources. In this process, 1-butene is isomerized to 2-butene, and the two are directly reacted during the metathesis process over a particular unique catalyst.

Brown Kellogg Company has developed a wonderful technology for producing propylene and ethylene required in the global market from cheap materials and oil residues of C_4 to C_8 fractions, especially olefinic fractions of C_4 and C_5, available in the ethylene plants and refineries light gases such as the products of catalytic cracking of naphtha and the out of specification motor gasolines.

Table 4-3: An ideal distribution of products from Superflex Process using different feeds.

Product/feed	FCC products	Coker	Pygas C_4	Pygas C_5
Gas oil	6.1	6.1	2.7	12
Ethylene	20	8.2	5.2	1.2
Propylene	40	7.4	2.5	8.4
Propane	6.6	7	3.5	5.6
Gasoline C_{6+}	7.3	9.2	8.2	6.2

This process is called Superflex and relies on a wonderful catalyst that does not require any treatment for catalyst poisons such as sulfur, water, oxygen or nitrogen compounds. The catalytic process converts paraffins and olefins into the required products, and the catalyst is regenerated continuously, which allows the use of high temperature. Table (4-3) summarizes an ideal distribution of the products using different feeds. Figure (4-1) illustrates the most important petrochemical derivatives of propylene.

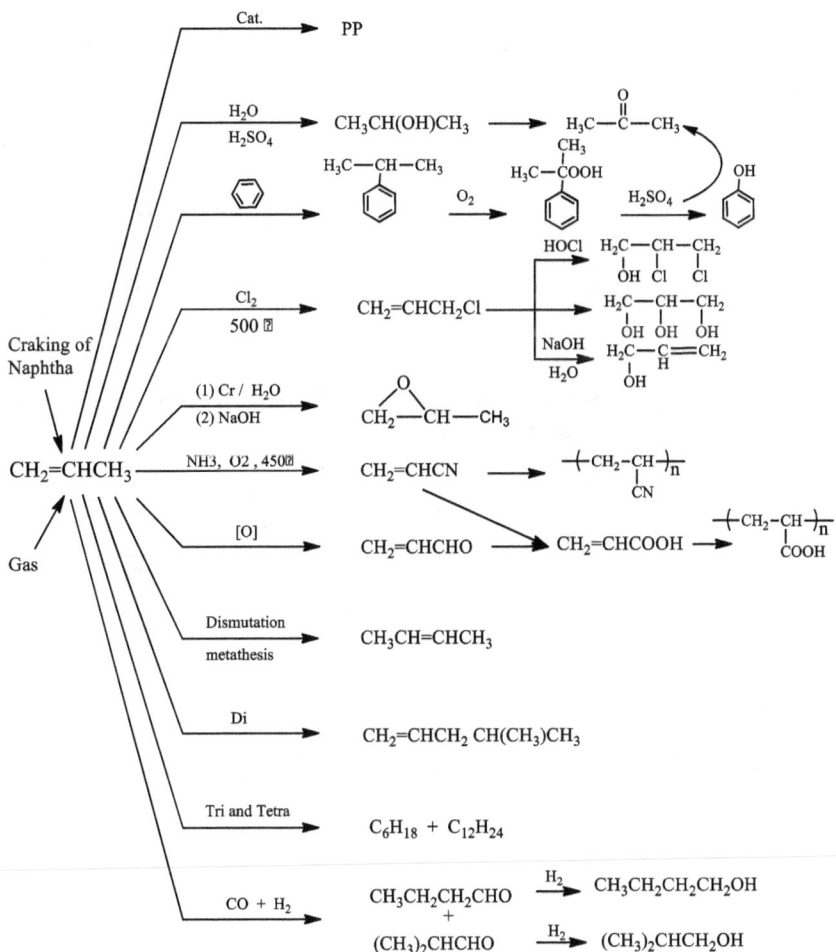

Figure 4-1: The most important petrochemical derivatives of propylene.

99

Polypropylene

Polypropylene is used in the production of many industrial final applications such as molding, bottoming, injection, industry of sheets and films, fibers, and pipes industry, in addition to some engineering applications. The reaction is performed according to the following equation in ideal conditions, at 80-100°C and a pressure of 50-60 atmospheres:

$$n \ CH_2=CHCH_3 \longrightarrow -\!\!\left(CH_2-CH\right)_n \ | \ CH_3$$

Table 4-4: The most significant differences and characteristics of the different polypropylene technologies.

Company	Total production mm ton/a	Flow rate	Production m ton/a	Technology
Basell*	5.13	—	40-400	Spherepol
Borealis	2.0	1-1200	200	Borstar
BP	—	—	65-350	BP Amoco
Chisso	5.1	—	65-300	Horizontal reactor
Mitsui	2.2	—	—	
Union Carbide		1-3000	26-80	

* Feed 94 % pure.

Polypropylene can be produced as a homogeneous polymer or as a copolymer, using famous co-monomers such as ethylene and butene. Ziegler-Natta catalysts were used to prepare isotactic polypropylene with an isotactic index in the range between 90 and 99%. Impact polypropylene is also produced together with other special polypropylenes to commensurate with the requirements of the final applications.

The modern generations of catalysts provided many wonderful products. Six international companies are competing in the market place in the polypropylene technologies. Table (4-4) summarizes the most important differences and characteristics of the different polypropylene technologies owned by those institutions.

Isopropyl alcohol

Organic chemistry students remember Markovnikov's rule where addition of water to a double bond gives secondary and tertiary alcohols. This process was used in the case of propylene for the production of isopropyl alcohol since 1930. It was the first example of the production of petrochemical products from refinery products.

Isopropyl alcohol is synthesized by hydration of propylene in three ways:

a) Indirect hydration in two steps by sulfuric acid in the liquid state:

$$CH_3CH{=}CH_2 + H_2SO_4 \longrightarrow (CH_3)_2CHO\overset{\overset{\displaystyle O}{||}}{\underset{\overset{||}{\displaystyle O}}{S}}OH \overset{H_2O}{\longrightarrow} CH_3\overset{\overset{\displaystyle CH_3}{|}}{C}HOH$$

This method suffers from severe corrosion caused by sulfuric acid, and large quantities of contaminated water, wastewater, and finally high air pollution.

b) Direct hydration in one step, catalyzed heterogeneously by an acidic ion exchange resin, in high-pressure and low temperature. The reaction is exothermic with reduction of the number of moles, according to Le Chatelier principle. This reaction occurs in high rate, due to the great stability of the isopropyl carbonium ion, compared to the ethylium ion for example, as in the hydration of ethylene:

$$CH_3CH=CH_2 + H_2O \longrightarrow CH_3\overset{\displaystyle CH_3}{\underset{\displaystyle |}{C}}HOH$$

c) Direct hydration in one step, called the Tokuyama method, can be catalyzed homogeneously in the liquid state at 270°C and 200 atmospheres, with a selectivity approaching 99% to isopropanol over tungsten silicate acids as catalysts.

Isopropyl alcohol is used as a solvent in extraction, and an alternative to ethanol in the cosmetics and pharmaceutical industry.

One of the most important derivatives of isopropyl alcohol is isopropyl acetate, which is synthesized by the esterification of isopropyl alcohol with acetic acid directly, in the presence of an acidic ion exchange catalyst.

The other important derivatives of isopropyl alcohol are isopropyl amine, which is used as an intermediate in industries such as paints, rubber, pesticides and isopropyl oleate required in the manufacture of cosmetics.

Acetone

Acetone is the simplest and the most essential industrial ketone. The World consumption of acetone exceeds five million tons annually. Acetone is manufactured industrially by three methods:

a) The first method of preparation of acetone is during the synthesis of phenol through Hawk cumene famous process, as a byproduct, which will be discussed elsewhere in this book.

102

b) The second method of synthesis of acetone is by direct oxidation of propylene, in Wacker-Hoechst process.

c) The third method is through dehydrogenation of isopropyl alcohol.

Acetone is produced in the fermentation processes in 10% yield, together with ethanol and n-butanol.

Acetone is also obtained through the oxidation processes of paraffins with oxygen, such as isobutane as 36% of the total product.

Perhaps the synthesis of acetone in Wacker-Hoechst process is the most convenient and popular. It is practiced in the industry since 1964. In the process, propylene is oxidized in the air with high selectivity in the liquid state at about 110°C, and 12 atmospheres of pressure, in the presence of an oxidation and reduction catalytic system containing palladium chloride:

$$CH_3CH=CH_2 + \tfrac{1}{2}O_2 \xrightarrow{\text{cat.}} CH_3\overset{\displaystyle CH_3}{\overset{\displaystyle |}{C}}=O$$

As in the case of oxidation of ethylene to acetaldehyde, in this system, palladium chloride is reduced to the palladium element and then re-oxidized to palladium chloride through a cupric chloride/cuprous chloride redox system.

Wacker method is used as well in converting normal butane into methyl ethyl ketone.

The cheap acetone product obtained as a byproduct in a number of industrial processes prevented the development of new technologies.

Acetone is synthesized from isopropyl alcohol in the gas phase by abstraction of hydrogen through oxidation in the presence of air or oxygen, at 500°C, under the action of silver or copper as a catalyst. Alternatively, acetone is synthesized by the direct dehydrogenation at 300°C using a catalyst of zinc oxide:

103

$$\underset{\text{CH}_3}{\underset{|}{\text{CH}_3\text{CHOH}}} + \tfrac{1}{2}\,O_2 \xrightarrow{\text{cat.}} \underset{\text{CH}_3}{\underset{|}{\text{CH}_3\text{C}{=}\text{O}}} + H_2O$$

$$\underset{\text{CH}_3}{\underset{|}{\text{CH}_3\text{CHOH}}} \xrightarrow{\text{cat.}} \underset{\text{CH}_3}{\underset{|}{\text{CH}_3\text{C}{=}\text{O}}} + H_2$$

In the case of liquid state, acetone is prepared from isopropyl alcohol through dehydrogenation in the presence of Raney nickel or copper chromate at 150°C at normal pressure.

In collaboration with DuPont, Shell has developed an auto-oxidation method for the preparation of acetone from isopropyl alcohol. The main purpose of this process is to obtain hydrogen peroxide. The free radical reaction is commenced at about 100°C, 3-4 atmospheres of pressure, using a few drops of hydrogen peroxide as an initiator as shown in the following equation:

$$\underset{\text{CH}_3}{\underset{|}{\text{CH}_3\text{CHOH}}} + O_2 \longrightarrow \underset{\text{CH}_3}{\underset{|}{\text{CH}_3\text{C}{=}\text{O}}} + H_2O_2$$

The following equations explain the mechanism of this reaction:

$$\underset{\text{CH}_3}{\underset{|}{\text{CH}_3\text{CHOH}}} + \overset{\bullet}{X} \longrightarrow \underset{\text{CH}_3}{\underset{|}{\text{CH}_3\underset{\bullet}{\text{C}}\text{OH}}} + HX$$

$$\xrightarrow{+ O_2} \quad \overset{O}{\overset{\|}{\text{CH}_3\text{CCH}_3}} + H\text{OO}^{\bullet}$$

$$H\text{OO}^{\bullet} + \underset{\text{CH}_3}{\underset{|}{\text{CH}_3\text{CHOH}}} \longrightarrow \underset{\text{CH}_3}{\underset{|}{\text{CH}_3\underset{\bullet}{\text{C}}\text{OH}}} + H_2O_2$$

Acetone can be used for the preparation of methyl methacrylate, methyl isobutyl ketone and bisphenol A, as the most principal products, despite the presence of other less essential products. Acetone products are used in the traditional reactions of ketones such as Aldol condensation and the formation of cyanohydrin.

Methyl isobutyl ketone

The most significant reactions of ketones using the Aldol condensation system give methyl isobutyl ketone, or 2-methyl-4-pentanone. Acetone proceeds in three steps to give the final product, provided that the separation of the intermediate compounds is possible, such as diacetone alcohol and mesityl oxide. These products are themselves essential.

In the first step of the reaction, acetone reacts with itself in the liquid state, to give diacetone alcohol:

$$
\underset{\substack{| \\ CH_3C=O}}{\overset{CH_3}{}} + \underset{\substack{| \\ CH_3C=O}}{\overset{CH_3}{}} \xrightarrow{OH^-} \underset{\substack{| \\ OH}}{\overset{CH_3 \quad O}{CH_3C-CH_2CCH_3}}
$$

In the second step, water is abstracted, in the presence of sulfuric acid as a catalyst or phosphoric acid at 100°C:

$$
\underset{\substack{| \\ OH}}{\overset{CH_3 \quad O}{CH_3C-CH_2CCH_3}} \longrightarrow CH_3C=CHCCH_3
$$

Mesityl oxide is hydrogenated to methyl isobutyl ketone, and the latter is hydrogenated once more to methyl isobutyl carbinol (4-methyl-2-pentanol), at 150-200°C, and 3-10 atmospheres of pressure, using copper or nickel as a catalyst:

$$
\underset{\substack{| \\ CH_3}}{\overset{\substack{CH_3\ O \\ |\ \ \ || }}{CH_3C=CHCCH_3}} \xrightarrow{H_2} \underset{\substack{|| \\ O}}{\overset{\substack{CH_3 \\ |}}{CH_3CHCH_2CCH_3}} \xrightarrow{H_2} \underset{\substack{| \\ OH}}{\overset{\substack{CH_3 \\ |}}{CH_3CHCH_2CHCH_3}}
$$

Diacetone alcohol is hydrogenated to hexylene glycol:

$$
\underset{\substack{| \\ OH}}{\overset{\substack{CH_3 \\ |}}{CH_3C}}\underset{}{-CH_2}\overset{\substack{O \\ ||}}{CCH_3} \xrightarrow{H_2} \underset{\substack{|\ \ \ | \\ OH\ OH}}{\overset{\substack{CH_3 \\ |}}{CH_3CCH_2CHCH_3}}
$$

Diacetone products are used as outstanding solvents in many industries, such as cellulose acetate, cellulose acetobutyrates, acrylic resins and alkyd resins. The same compounds are used also for the extraction of inorganic salts and organic compounds.

Acrylic acid and its esters

Acrylic acid and its esters are considered the most significant unsaturated carboxylic acids used in the industry. Furthermore, they are among the oldest compounds that found early industrial applications. Their industrial preparation has begun in Germany since 1901. A number of technologies have been developed for the synthesis of acrylic acid; however, the most notable is the oxidation of propylene:

$$
CH_3CH=CH_2 + O_2 \xrightarrow{cat.} CH_2=CHCHO \xrightarrow{cat.\ \frac{1}{2}\ O_2} CH_2=CHCOOH
$$

This process can be performed in one or two steps: when using a multifunctional catalyst, a mixture of acrolein and acrylic acid is produced in a single step, from the reaction of oxygen with propylene, at 200-500°C and 10 atmospheres of pressure, depending on the type of catalyst used. The polyfunctional catalysts used

106

typically contain molybdates of heavy elements, in addition to tellurium compounds as activators.

The two-step process is performed using different catalysts, resulting in acrolein in the first step, and acrylic acid in the second.

There are a number of traditional outdated methods of synthesis of acrylic acid that were disqualified from industrial competitiveness. These are summed up in the following paragraphs.

The first old method of synthesis of acrylic acid starts from the reaction of ethylene oxide with hydrogen cyanide in the presence of a base, to give hydroxypropyl nitrile. This product reacts with alcohol, or water, and an equivalent quantity of sulfuric acid, to give the ester, or acrylic acid, respectively:

$$H_2C \overset{O}{\underset{}{\triangle}} CH_2 + HCN \longrightarrow HOCH_2CH_2CN$$

$$HOCH_2CH_2CN + ROH + H_2SO_4 \longrightarrow H_2C{=}CHCOOR + NH_4HSO_4$$

The second old method is Reppe's method, where the carbonyl is added to acetylene together with water, or alcohol, in the presence of zinc carbonyl to provide the acid or the ester:

$$HC{\equiv}CH + CO + ROH \xrightarrow{\text{cat.}} CH_2{=}CHCOOR$$

The third old method is the ketene's method. The ketene is reacted with formaldehyde, in the presence of a Lewis acid, such as aluminum chloride as a catalyst, to give propiolactone, which disintegrates thermally, to produce the acid, or reacts with the alcohol to give the ester:

$$CH_2{=}C{=}O + CH_2{=}O \longrightarrow \underset{H_2C-O}{\overset{H_2C}{\big|}}{\overset{O}{\diagup}} \xrightarrow{\text{ROH}} CH_2{=}CHCOOR$$

107

The following traditional method is based on the hydration of acrylonitrile followed by esterification:

$$CH_2=CHCN + H_2SO_4 + H_2O \longrightarrow CH_2=CHCONH_2.H_2SO_4 + ROH$$

$$\longrightarrow CH_2=CHCOOR + NH_4HSO_4$$

It is clear that the biggest challenge of this technology is to obtain an equivalent quantity of ammonium bisulphate, with each equivalent of acrylic acid.

The esters of acrylic acid are prepared from the acid. The most significant esters are methyl acrylate, ethyl acrylate, butyl acrylates; n- and iso-octyl (2-ethylhexane) acrylates. All the esters are polymerized individually or as co-monomers and used in paint industries, stickers, paper, fabrics, and in treatment of leather. Esterification processes take place in the presence of an acidic ion exchange catalyst, or through trans-esterification of methyl acrylate.

Methacrylic acid and its esters

The significance of methacrylic acid is in its esters such as methyl ester, which is the most prominent. Methacrylate esters are used in the production of polymers and copolymers.

The oldest traditional method in the industrial field that prevailed since 1937 until today for the synthesis of acrylic acid and its derivatives starts from acetone cyanohydrin. It holds more than 80% of the volume of the global production. The operation is carried out in two steps: first, addition of hydrogen cyanide to acetone, in the presence of a catalytic base at mild conditions:

In the second step, the nitrile group is hydrated with concentrated sulfuric acid (98%) at about 100°C, to give the amide as an intermediate. The latter is converted directly into the methyl ester, by the reaction with methanol, at about 100°C:

$$\underset{\substack{HO \quad CN}}{\overset{CH_3}{\underset{|}{C}}CH_3} \longrightarrow \underset{\substack{HO \quad CNH_2 \cdot H_2SO_4 \\ \| \\ O}}{\overset{CH_3}{\underset{|}{C}}CH_3} \xrightarrow{+ CH_3OH} \underset{\substack{COCH_3 \\ \| \\ O}}{CH_2=C}\overset{CH_3}{} + NH_4HSO_4 + H_2O$$

Mitsubishi Gas Chemical Company developed the aforementioned method, to avoid the formation of mono ammonium sulfate as a byproduct. It adopted the partial hydrolysis of acetone cyanohydrin to the alpha-hydroxyisobutyramide, which was reacted with methyl formate, to give the methyl ester and the formamide:

$$\underset{\substack{HO \quad CN}}{\overset{CH_3}{\underset{|}{C}}CH_3} \xrightarrow{H_2O} \underset{\substack{HO \quad CNH_2 \\ \| \\ O}}{\overset{CH_3}{\underset{|}{C}}CH_3} \xrightarrow{HCOOCH_3} \underset{\substack{COCH_3 \\ \| \\ O}}{CH_2=C}\overset{CH_3}{} + HCONH_2$$

With this process the outcome of this reaction becomes as follows:

$$\underset{\substack{HO \quad CN}}{\overset{CH_3}{\underset{|}{C}}CH_3} + HCOOCH_3 \longrightarrow \underset{\substack{COOCH_3}}{CH_2=C}\overset{CH_3}{} + HCONH_2 + H_2O$$

The produced formamide is utilized to produce hydrogen cyanide, to react with acetone, to yield cyanohydrin acetone.

$$\underset{|}{\overset{CH_3}{CH_3C=O}} + HCN \longrightarrow \underset{\substack{HO \quad CN}}{\overset{CH_3}{\underset{|}{C}}CH_3}$$

Many attempts were made for the synthesis of methacrylic acid from cheap raw materials, such as propionaldehyde, isobutene, isobutyraldehyde, as well as the desire to avoid the formation of ammonium sulfate, which accompanies the cyanohydrin method.

BASF company has developed a new methacrylic acid method of synthesis starting from formaldehyde and propionaldehyde that is produced from the reaction of ethylene with formaldehyde, in the presence of a secondary amine and acetic acid, at 160-210°C, and a pressure of 60 atmospheres, to give methacrolein. This in turn is oxidized to methacrylic acid that is esterified to give methyl methacrylate:

$$CH_3CH_2CHO + HCHO \xrightarrow{\text{[sec. amine]} + H^+} CH_2=C\begin{smallmatrix}CH_3\\\\CH\\||\\O\end{smallmatrix} + H_2O \xrightarrow{+ O_2} CH_2=C\begin{smallmatrix}CH_3\\\\COH\\||\\O\end{smallmatrix}$$

Scampia company began a process for developing this synthesis from isobutene, in two steps: an oxidation process using nitrogen dioxide where alpha hydroxy isobutyric acid is formed first, which expels water, giving methacrylic acid. However, an explosion happened during the industrial operation of this process, which necessitated abandoning it:

$$CH_2=C\begin{smallmatrix}CH_3\\\\CH_3\end{smallmatrix} \xrightarrow{N_2O_4} \underset{HO}{\overset{CH_3}{\underset{\;}{C}}}\begin{smallmatrix}\\CH_3\\COOH\end{smallmatrix} \xrightarrow{-H_2O} CH_2=C\begin{smallmatrix}CH_3\\\\COOH\end{smallmatrix}$$

A two-step process has been developed in Japan for the synthesis of methacrylic acid where an intermediate tertiary butanol is firstly obtained from a mixture of n-butanes and butenes in the liquid state in the presence of a catalyst of an acid ion exchange resin. The second step is a heterogeneous oxidation by air to methacrolein at 420°C, and atmospheric pressure or slightly higher:

$$CH_2{=}C\underset{CH_3}{\overset{CH_3}{\diagup}} \quad \xrightarrow{+[H^+]+H_2O} \quad HO{-}C\underset{CH_3}{\overset{CH_3}{\diagup}}{-}CH_3 \quad \xrightarrow{+[cat.]\ +O_2\ -2H_2O} \quad CH_2{=}C\underset{CHO}{\overset{CH_3}{\diagup}}$$

Methacrolein is oxidized in the presence of water vapor, at 300°C, and a pressure of 2-3 atmospheres, over a catalyst of oxides of molybdenum, phosphorus and vanadium:

$$CH_2{=}C\underset{\underset{O}{\overset{\|}{CH}}}{\overset{CH_3}{\diagup}} \quad \xrightarrow{+\ O_2} \quad CH_2{=}C\underset{\underset{O}{\overset{\|}{COH}}}{\overset{CH_3}{\diagup}}$$

The third cheap feed used in the synthesis of methyl methacrylate is isobutyraldehyde. It is oxidized in the first step in the presence of a heterogeneous catalyst, using hydrogen bromide, at 170°C, to isobutyric acid from which hydrogen is stripped to give methacrylic acid:

$$\underset{H}{\overset{CH_3}{\diagup}}C\underset{CHO}{\overset{\vphantom{|}}{\diagdown}}CH_3 \quad \xrightarrow{HBr} \quad \underset{H}{\overset{CH_3}{\diagup}}C\underset{COOH}{\overset{CH_3}{\diagdown}}CH_3 \quad \xrightarrow{-H_2} \quad CH_2{=}C\underset{COOH}{\overset{CH_3}{\diagup}}$$

Methacrylic acid is used for the manufacture of Plexiglas, the transparent crystalline plastic with extraordinary specifications of firmness, resistance to breakage, and chemical stability. Also from methacrylic acid, many copolymers were synthesized with a broad range of applications.

Acrolein

The traditional method of synthesis of acrolein starts from the condensation of acetaldehyde with formaldehyde, at 300°C, in the gaseous state, over a catalyst of sodium on silica. This method is known as Degussa method. It is in service since 1942:

$$CH_3CHO + HCHO \xrightarrow{\text{cat.}} CH_2=CHCHO + H_2O$$

Figure (4-2) shows the main uses of acrolein.

Recently, acrolein is synthesized from propylene, as an available and cheap starting material, by oxidation at 350-400°C, under a pressure of 1-2 atmospheres, in the presence of a catalyst of cuprous oxide on silica and iodine as an activator. A number of catalysts have been introduced to develop this reaction such as the oxides of bismuth and molybdenum among others:

$$CH_3CH=CH_2 + O_2 \xrightarrow{\text{cat.}} CH_2=CHCHO + H_2O$$

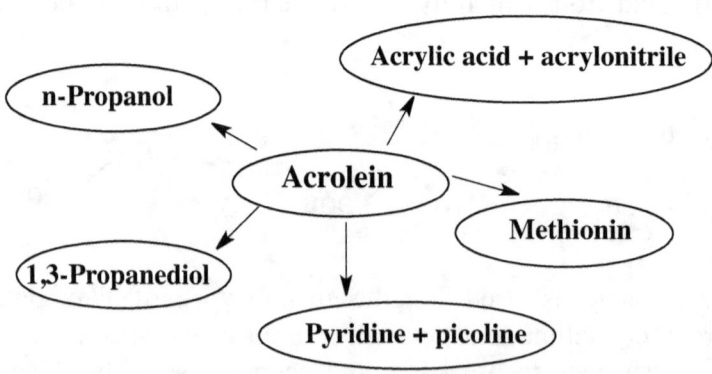

Figure 4-2: The main uses of acrolein.

Propylene oxide

The global demand for propylene oxide is constantly growing despite the manufacture difficulties. All the attempts made to oxidize propylene directly in the presence of silver, or another metal as a catalyst, as is the case of the synthesis of ethylene oxide, have simply failed. The preparation of propylene oxide globally takes place in two ways only: the chlorohydrin method and the indirect method of oxidation.

Propylene oxide is synthesized in two steps: Addition of chlorohydrin, then stripping hydrochloric acid. The chlorohydrin is characterized by preferential addition of Cl^+ on the terminal carbon of propylene according to Markovnikov's rule:

$$2\ CH_3CH=CH_2 + 2\ HOCl \longrightarrow H_3C-\underset{\underset{\displaystyle OH}{|}}{\overset{\overset{\displaystyle H}{|}}{C}}-\underset{\underset{\displaystyle Cl}{|}}{CH_2} + H_3C-\underset{\underset{\displaystyle Cl}{|}}{\overset{\overset{\displaystyle H}{|}}{C}}-\underset{\underset{\displaystyle OH}{|}}{CH_2}$$

$$\xrightarrow{2\ Ca(OH)_2} H_3C-\overset{\displaystyle O}{\overset{\displaystyle \diagup\diagdown}{\underset{\underset{\displaystyle H}{|}}{C}}}-CH_2 + Ca(Cl)_2 + 2H_2O$$

This process suffers from the conversion of chlorine to calcium chloride, which if not discharged appropriately would cause a big problem. Addition of chlorine to the double bond of propylene gives 1,2-dichloropropylene as a byproduct in addition to large quantities of contaminated water.

The indirect method of oxidation of propylene takes place through free radical oxidation, to give an oxygen atom from the peroxide formed to propylene, as in the following equation, in the case of using hydroperoxide:

$$CH_3CH=CH_2 + ROOH \xrightarrow{cat.} H_3C-\overset{\displaystyle O}{\overset{\displaystyle \diagup\diagdown}{\underset{\underset{\displaystyle H}{|}}{C}}}-CH_2 + ROH$$

The reaction might also use acid peroxide as follows:

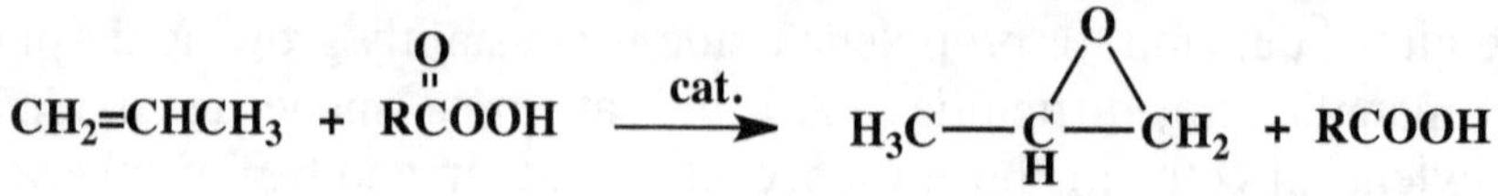

$$CH_2=CHCH_3 + \overset{O}{\overset{\|}{RCOOH}} \xrightarrow{\text{cat.}} H_3C-\underset{H}{\overset{O}{C}}-CH_2 + RCOOH$$

The process has been prepared easily at the same conditions through peroxy metals such as molybdenum as discussed previously.

In the liquid state, the process proceeds from isobutane by using t-butanol as a solvent, in the presence of a catalyst of molybdenum compounds, at about 100°C, and a pressure range between 15 and 65 atmospheres:

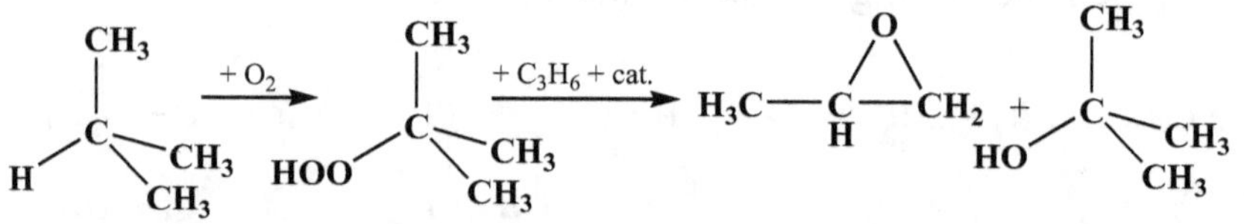

t-Butanol is produced from isobutane in addition to propylene oxide, which are separated by distillation. t-Butanol is added to motor gasoline to prevent freezing, and raise the octane number. Water can be removed from t-butanol to get isobutylene or converted to acrylic acid through Mitsubishi's modern silk technology.

Among the acids used in this process is acetic acid. Its peroxide is prepared by auto-oxidation of acetaldehyde as mentioned earlier, or through the reaction between acetic acid and hydrogen peroxide as follows:

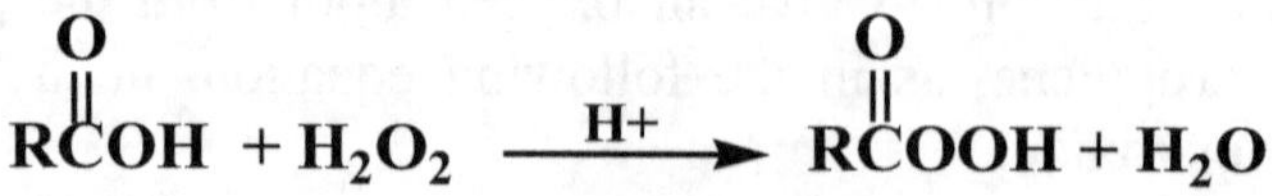

$$\overset{O}{\overset{\|}{RCOH}} + H_2O_2 \xrightarrow{H+} \overset{O}{\overset{\|}{RCOOH}} + H_2O$$

Allyl chloride

The importance of allyl chloride stems from its use as a source for the production of allyl compounds. Allyl chloride is produced from propylene by the addition of chlorine at the allyl carbon, at temperatures exceeding 300°C. The reaction becomes quantitative at 500-510°C, with more than 85% selectively:

$$CH_3CH=CH_2 + Cl_2 \longrightarrow ClCH_2CH=CH_2 + HCl$$

The reaction necessitates the use of an excess of propylene for chlorine to give the required product with mixing the reactants well before entering the reaction area. The hydrogen chloride formed is washed with water.

Most of the allyl chloride produced is used in the synthesis of epichlorohydrin (about 90%), allyl alcohol and allyl amine.

The process of synthesis of epichlorohydrin takes place in two steps: firstly, the reaction of allyl chloride with chlorohydrin, at room temperature (25-30°C) in the aqueous phase, to give a mixture of dichlorohydroxypropane:

$$2\ CH_2CH=CH_2 + 2\ HOCl \longrightarrow H_2C-\underset{Cl}{\overset{H}{C}}-CH_2 + H_2C-\overset{H}{C}-CH_2$$
(with CH below CHCl / Cl OH Cl and Cl Cl OH)

The raw product transforms into epichlorohydrin in the second step in the presence of calcium hydroxide at 50-90°C:

$$H_2C-\overset{H}{C}-CH_2 \xrightarrow{Ca(OH)_2} H_2C-C-CH_2 + Ca(Cl)_2 + 2H_2O$$
(Cl Cl OH → epoxide with O, H, Cl)

The oxidation of allyl chloride to epichlorohydrin is suggested to avoid the formation of calcium chloride salt using peroxypropionic acid, which is prepared from the reaction of propionic acid with hydrogen peroxide at 70-80°C:

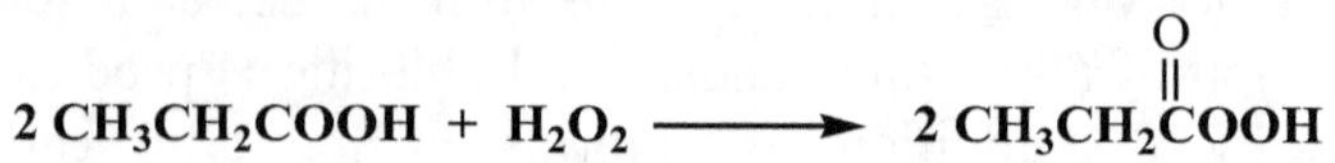

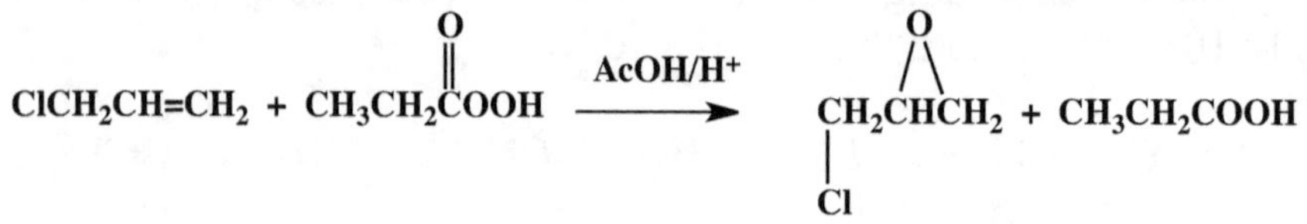

$$2\ CH_3CH_2COOH + H_2O_2 \longrightarrow 2\ CH_3CH_2\overset{\displaystyle O}{\overset{\|}{C}}OOH$$

$$ClCH_2CH{=}CH_2 + CH_3CH_2\overset{\displaystyle O}{\overset{\|}{C}}OOH \xrightarrow{\text{AcOH/H}^+} \underset{\underset{\displaystyle Cl}{|}}{CH_2}\overset{\displaystyle\overset{O}{\diagup\diagdown}}{CH}CH_2 + CH_3CH_2COOH$$

Epichlorohydrin is used in the preparation of ether chlorohydrins with bisphenol A in the presence of caustic soda. This reaction has been discussed elsewhere in this book.

The reaction of allyl chloride with ammonia gives allyl amine:

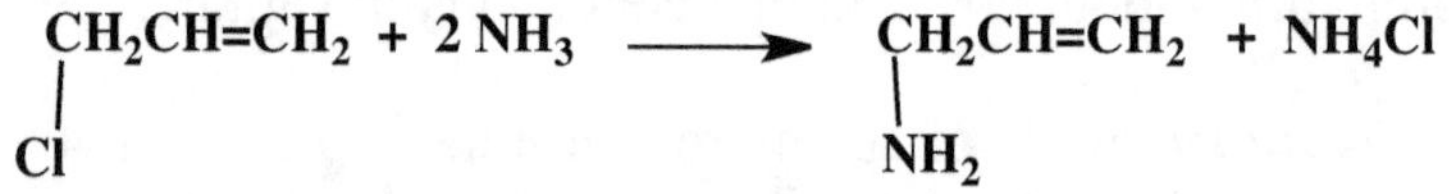

$$\underset{\underset{\displaystyle Cl}{|}}{CH_2}CH{=}CH_2 + 2\ NH_3 \longrightarrow \underset{\underset{\displaystyle NH_2}{|}}{CH_2}CH{=}CH_2 + NH_4Cl$$

Allyl alcohol

Allyl alcohol is synthesized industrially in four ways, starting from allyl chloride, propylene oxide, acrolein and allyl acetate. The most favorable method is the reaction of allyl chloride with caustic soda:

$$ClCH_2CH{=}CH_2 + NaOH \longrightarrow HOCH_2CH{=}CH_2 + NaCl$$

The final loss of chlorine and expensive industrial equipment needed to prevent corrosion are considered the main obstacles in this process.

The synthesis of allyl alcohol by isomerization of propylene oxide has been met with a boom due to the availability of the latter. The

116

process takes place in the presence of lithium phosphate as a catalyst, in the gaseous state, or the liquid state at 275°C and a pressure of 10 atmospheres, in a high boiling solvent:

$$\underset{\underset{H}{|}}{H_3C} - \overset{\overset{O}{\triangle}}{\underset{|}{C}} - CH_2 \longrightarrow HOCH_2CH=CH_2$$

The synthesis of allyl alcohol takes place by the hydrogenation of acrolein in the gaseous state. Alternatively, an excellent substitute for hydrogenation is the replacement of isopropyl alcohol or 2-butanol, according to the need of byproduct, whether acetone, or methyl ethyl ketone, at 400°C, in the presence of a catalyst of magnesium oxide and zinc oxide:

$$CH_2=CHCHO + CH_3\overset{CH_3}{\underset{|}{C}}HOH \longrightarrow HOCH_2CH=CH_2 + CH_3\overset{CH_3}{\underset{|}{C}}=O$$

$$CH_2=CHCHO + CH_3CH_2\overset{CH_3}{\underset{|}{C}}HOH \longrightarrow HOCH_2CH=CH_2 + CH_3CH_2\overset{CH_3}{\underset{|}{C}}=O$$

In this fine method of synthesis of allyl alcohol, propylene is reacted with acetic acid and oxygen in the presence of a catalyst of palladium to give the allyl acetate, that is hydrated to allyl alcohol, in the presence of an acidic catalyst, or an acidic ion exchange resin, at 230°C, and a pressure of 30 atmospheres, without a catalyst. The acetic acid produced as a byproduct is used again in the synthesis of allyl acetate:

$$CH_3CH=CH_2 + CH_3\overset{O}{\overset{||}{C}}OH + \tfrac{1}{2} O_2 \longrightarrow CH_3\overset{O}{\overset{||}{C}}OCH_2CH=CH_2 + H_2O$$

$$CH_3\overset{O}{\overset{||}{C}}OCH_2CH=CH_2 + H_2O \longrightarrow HOCH_2CH=CH_2 + CH_3\overset{O}{\overset{||}{C}}OH$$

117

Thus, the outcome of the reaction becomes oxidation of propylene, as follows:

$$CH_3CH=CH_2 + \tfrac{1}{2}\,O_2 \longrightarrow HOCH_2CH=CH_2$$

Glycerin

The entire world derived all its needs of glycerin from the saponification of fats as a highly demanded and useful byproduct in the manufacture of soap. However, the admission of synthetic detergents in the global market for nearly seventy years has reduced the saponification of natural fat, thus producing less natural glycerin; hence the intention of its synthesis from petrochemicals emerged.

Perhaps the best pathways in the synthesis of glycerin start from epichlorohydrin as a feedstock that is hydrated in the presence of dilute caustic soda in an aqueous reaction in two steps, at high pressure and 100-200°C:

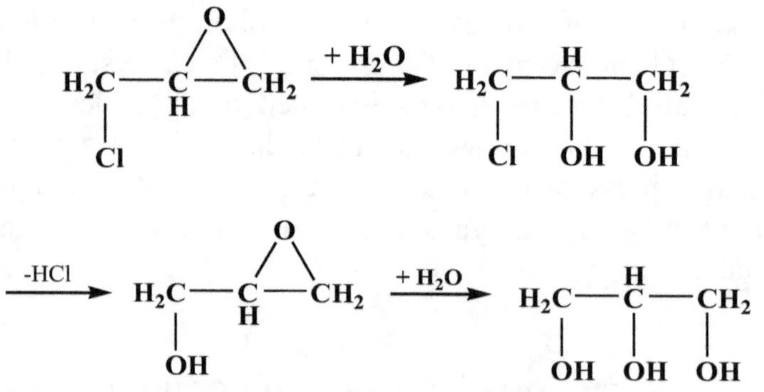

Glycerin was synthesized from allyl alcohol by reaction with hydrogen peroxide in the liquid state at 60-70°C in the presence of a catalyst of oxides or salts of tungstic acid:

118

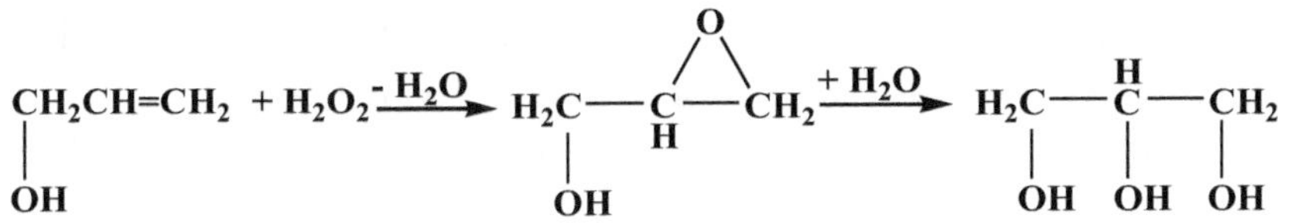

Some companies have replaced hydrogen peroxide with peroxyacetic acid, in a high boiling solvent, at 60°C.

Among the developments made to this process, after the oxidation of propylene oxide, it is reacted with HOCl in the second step to give epichlorohydrin, which is hydrated to glycerin. This method is preferred because it reduces the ratio of calcium chloride formed by half, compared with starting from allyl chloride:

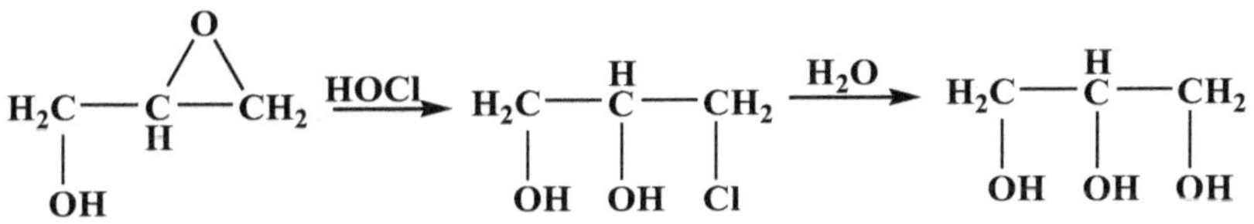

The uses of glycerin rely on its natural properties such as high boiling point, excessive viscosity, and absorption of water that make it suitable for use in medicinal industries, cosmetics, and as an anti-freeze agent. The chemical properties as a trihydroxyl alcohol qualify glycerin to be used in the synthesis of alkyd resins such as glyptal, an additive for networking in linear polyesters, the reactions of polyether isocyanates, and in the manufacture of paint and varnish.

Acrylonitrile

Acrylonitrile provides acrylic polymers with distinctive properties, whether alone or in the form of copolymers. One of these copolymers is acrylonitrile with butadiene, which gives a distinctive artificial rubber and other chemicals, intermediates, fibers, resins and thermoplastics.

Amino-oxidation of propylene is the most important method of synthesis of acrylonitrile in the modern era. It is prepared through the catalytic oxidation of ammonia, in the presence of oxygen to form the nitrile group:

$$CH_3CH=CH_2 + NH_3 + 1.5\,O_2 \longrightarrow CH_2=CHCN + 3H_2O$$

The chemical industry knew other traditional methods of synthesis of acrylonitrile from acetylene, ethylene oxide, acetaldehyde and propylene. These methods are summarized in the following equations:

$$H_2C \overset{O}{\overset{\diagup\diagdown}{-\!\!\!-}} CH_2 \xrightarrow{\;HCN\;} HOCH_2CH_2CN \xrightarrow{\;-H_2O\;} CH_2=CHCN$$

$$HC\equiv CH \;+\; HCN \longrightarrow H_2C=CHCN$$

$$CH_3CHO + HCN \longrightarrow \underset{\quad}{CH_3\overset{CN}{\overset{|}{C}}HOH} \xrightarrow{\;-H_2O\;} CH_2=CHCN$$

$$4CH_3CH=CH_2 + 6NO \longrightarrow 4CH_2=CHCN + 6H_2O + N_2$$

The most important uses of acrylonitrile are acrylic fibers, plastics, polyacrylonitrile-butadiene-styrene plastics, polyacrylonitrile-styrene plastics, nitrile rubber, adiponitrile and other additional products; the best known of them is acrylamide.

Di-, tri- and tetrapropylene

The synthesis of di-propylene leads to a number of hexenes, which vary in composition depending on the reaction conditions and type of catalyst used. A catalyst of potassium permanganate, supported by potassium carbonate, gives 4-methyl-1-pentene in 85% yield. The

120

following mechanism has been proposed to explain the course of the reaction:

$$CH_2=CH-CH_3 \xrightarrow{K} CH_2=CH^-CH_2^- \ K^+ \xrightarrow{CH_3-CH=CH_2}$$

$$CH_2=CH-CH_2-CH(CH_3)-CH_2^- \ K^+ \xrightarrow{CH_3-CH=CH_2}$$

$$K^+ \ ^-CH_2-CH=CH_2 + CH_2=CH-CH_2-\overset{\overset{\displaystyle CH_3}{|}}{C}H-CH_3$$

The product can be polymerized in the presence of a Ziegler-Natta catalyst to give poly(4-methyl-1-pentene), which is characterized by transparency, good resistance to heat and chemicals and used in the manufacture of medical and laboratory instruments and equipment.

When using a catalyst of tripropyl aluminum, at 200°C, and 200 atmospheric pressures, the product achieved was 2-methyl-1-pentene that was isomerized at 200°C, on a catalyst of industrial zeolite to 2-methyl-2-pentene. This product is converted to isoprene when passed in a pyrolysis oven with steam, in the presence of hydrogen bromide as a catalyst:

$$CH_2=CH-CH_3 + CH_2=CH-CH_3 \longrightarrow CH_2=CH-CH_2-\overset{\overset{\displaystyle CH_3}{|}}{C}H-CH_3$$

$$\longrightarrow CH_3CH_2CH=\overset{\overset{\displaystyle CH_3}{|}}{C}-CH_3 \longrightarrow CH_2=CH\overset{\overset{\displaystyle CH_3}{|}}{C}=CH_2 + CH_4$$

Perhaps it is appropriate to recall here that the alkyl benzene used in the synthesis of sodium sulfonate detergent was industrially prepared by the alkylation of benzene using a mixture of C_{12} alkene (four propylenes). The latter is synthesized together with a tri-propylene by the reaction of propylene with itself for three or four times in the presence of an acid catalyst:

$$CH_2=CH-CH_3 \xrightarrow{H^+} CH_2-{}^+CH-CH_3 \xrightarrow{CH_2=CH-CH_3}$$

$$\underset{\underset{CH_3}{|}}{CH_3-CH-CH_2-{}^+CH-CH_3} \xrightarrow{CH_2=CH-CH_3} C_9H_{19}{}^+ \xrightarrow{CH_2=CH-CH_3} C_{12}H_{25}{}^+$$

$$C_9H_{19}{}^+ \downarrow {-H^+}$$
$$C_9H_{18}$$

$$C_{12}H_{25}{}^+ \downarrow {-H^+}$$
$$C_{12}H_{24}$$

The use of alkyl benzene in the detergent industry has declined due to the existence of many branches in the alkyl group. Branching does not allow biodegradation, leading to the contamination of water resources and the formation of foam in the sewerage systems. The branched alkyl chains have been replaced by straight chain ones that are biodegradable.

Tri-propylene is used in the synthesis of non-ionic detergents through reaction with phenol:

Nonylphenol

Chapter Five
Petrochemicals from Butenes and Butadiene

Butenes 124
Higher olefins 125
Branched higher olefins 127
Butadiene 129
Isoprene 133
Butan-1,4-diol 134
1-Butene 137
n- and i-Butyraldehyde 138
Butyl alcohols 139
Vinyl ethers 141

Butenes

Butenes are extracted as byproducts from the refining operations, the cracking of hydrocarbons and through the condensation of acetylene with itself.

Butenes can be obtained industrially in abundance from propylene through disproportional cracking. Concisely, this happens as follows:

$$2 \ CH_2=CHCH_3 \longrightarrow CH_2=CH_2 + CH_3\text{-}CH=CH\text{-}CH_3$$

However, the most important sources of C_4 are the fluid catalytic cracking of diesel and the steam cracking of hydrocarbon compounds from ethylene to diesel. The most valuable source of carbon C_4 fraction is the steam cracking of naphtha.

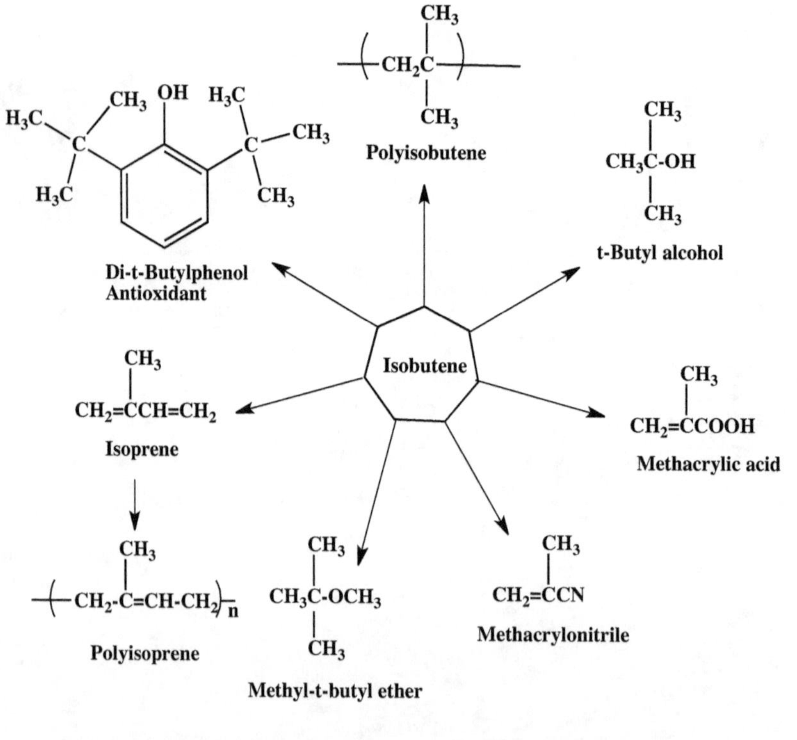

Figure 5-1: The most important industrial uses of isobutene.

124

Table (5-1) summarizes the composition of C_4 fraction produced from the steam cracking of naphtha and the catalytic cracking of diesel (weight %).

It is noted from Table (5-1) that with the temperature increase the C_4 proportion reduced, but the proportion of butadiene increased. This is the same trend of increase of ethylene. Butadiene is stable even at high temperatures, because of its characteristic resonance energy. Figure (5-1) summarizes the most important uses of isobutene.

Table 5-1: The composition of C_4 fraction produced from the steam cracking of naphtha and the catalytic cracking of diesel (weight %).

Cracking product	Steam cracking		Catalytic cracking (Zeolite)
	Low temperature	High temperature	
1,3-Butadiene	26	47	5
Isobutene	32	22	15
1-Butene	20	14	12
Trans-2-butene	7	6	12
Cis-2-butene	7	5	11
Butane	4	3	13
Isobutane	2	1	32
Other butanes	2	2	-

Higher olefins

The long chains of the olefins range between C_5 and C_{20}. Of course, they can be a series of straight or branched chains. The double bond can be terminal or internal. Each of these higher olefins has its own characteristic properties and specifications and method of synthesis and specific applications. In the thermal cracking conditions, hydrogens are abstracted from the terminal carbons, and alpha olefins are obtained. In the circumstances of the catalytic cracking conditions, the hydrogens are abstracted from the middle of the chains and internal olefins are obtained. In the case of the desire to obtain linear internal olefins, n-paraffins are chlorinated followed by catalytic abstraction of hydrogen chloride, at 250-350°C, over aluminum silicate as a catalyst:

$$R^1CH_2CH_2R^2 \xrightarrow{+\ Cl_2} R^1\overset{Cl}{\overset{|}{CH}}\text{-}CH_2R^2 + R^1CH_2\text{-}\overset{Cl}{\overset{|}{CH}}R^2 \xrightarrow{-HCl} R^1CH=CHR^2$$

Linear olefins in the range C_6-C_{10} are utilized, through hydroformylation, followed by hydrogenation, to prepare alcohols that are exploited themselves as plasticizers, or in the manufacture of plasticizers. Internal olefins and alpha olefins produce similar mixtures of aldehydes after hydroformylation because of the instant isomerization of the double bond:

$$CH_3(CH_2)_3CH=CH_2 + HCHO \xrightarrow{H_2} CH_3(CH_2)_5CH_2OH + CH_3(CH_2)_4\overset{OH}{\overset{|}{CH}}CH_3$$

The C_{10} to C_{13} olefin chains are used in the alkylation of benzene to produce linear alkyl benzene as the first step in the production of synthetic detergents.

$$CH_3(CH_2)_{10}CH=CH_2 + \text{⬡} \longrightarrow \text{⬡}\!-\!C_{13}H_{27}$$

The olefin chains of C_{14}-C_{18} can be directly reacted with sulfur trioxide for the production of the sulfonates, which are used as synthetic detergents.

$$CH_3(CH_2)_{12}CH=CH_2 + SO_3 \longrightarrow CH_3(CH_2)_{12}CH_2CH_2SO_2OH$$

The hydration of the ring with water is essential to increase the solubility of the product. The alkyl sulfonates are considered effective and inexpensive detergents, and can be easily synthesized from n-paraffins, ranging between C_{12} and C_{18}, through chlorosulfonation, and oxosulfonation:

$$R^1CH_2R^2 + SO_2 + Cl_2 \xrightarrow{hv\ -HCl} R_1\overset{\overset{\displaystyle R^2}{|}}{C}HSO_2Cl \xrightarrow{+2NaOH} R_1\overset{\overset{\displaystyle R^2}{|}}{C}HSO_3Na$$

$$R^1CH_2R^2 + SO_2 + \tfrac{1}{2}O_2 \longrightarrow R^1\overset{\overset{\displaystyle R^2}{|}}{C}HSO_2OH$$

Branched higher olefins

Due to the industrial need for branched mono olefins in various areas they have been synthesized by oligomerization, and co-oligomerization of propylene, isobutene and n-butenes. Table (5-2) summarizes the raw materials and type of process used for producing the branched olefins and the respective products. It is noticeable that three types of catalysts identified the anticipated products, namely:

1) Protonic catalysts such as phosphoric acid on silica (H_3PO_4/SiO_2) are used for the oligomerization of propylene to give the branched olefins, which are hydrogenated then added to motor gasoline, to raise the octane number.

2) Sulfuric acid is used as well to catalyze the oligomerization of isobutene, to give excellent additives to motor gasoline, and iso-nonanol which is used in the manufacture of alcohols for plasticizers:

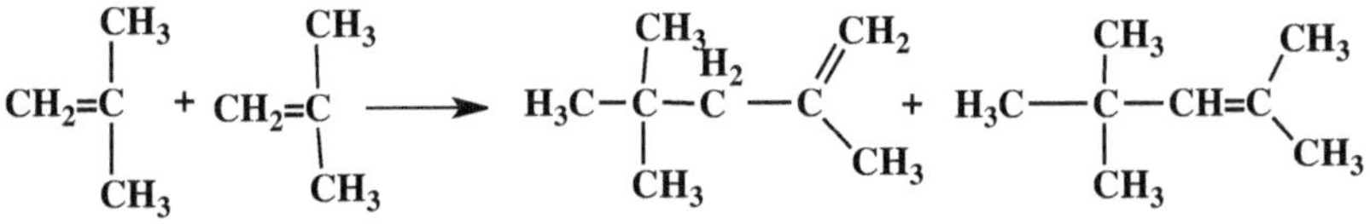

This process is suitable for propylene in addition to butene.

3) The alkyl aluminum catalysts process requires conditions of 200°C and 200 atmospheres, in the presence of a co-catalyst for better selectivity. Without the co-catalyst the catalyst is less effective. Upon addition of the transition metals compounds or elements or their complexes, the effectiveness increases and selectivity decreases. As an example of this, the following equation is included for the preparation of 2-methylpentene, which is used as feedstock to prepare isoprene:

$$CH_2=CH-CH_3 + CH_2=CH-CH_3 \xrightarrow{AlPr_3} \overset{\overset{\displaystyle CH_3}{|}}{CH_2=CCH_2CH_2CH_3}$$

Upon addition of a co-catalyst of nickel salts at 60°C, and 18 atmospheres, the reaction of propylene with butene becomes highly selective, and produces iso-heptanes with relatively few branches. At the same time, propylene reacts with itself, and butene reacts with itself too. Propylene itself oligomerizes, as well as butene, to give high selectivity products with fewer branches. The iso-olefin chains of C_6, C_7 and C_8, with little branching, are very suitable for hydroformylation and hydrogenation, to give iso alcohols, which are used in the manufacture of dialkyl phthalate plasticizers.

4) Alkali metal catalysts can be used to dimerize propylene, in the liquid state, to 4-methyl-1-pentene in high selectively, and low efficiency, as in the following British Petroleum Company technology:

$$CH_2=CH-CH_3 + CH_2=CH-CH_3 \xrightarrow{Na/K_2CO_3} \overset{\overset{\displaystyle CH_3}{|}}{CH_2=CHCH_2CHCH_3}$$

Table 5-2: Raw materials and type of process for producing branched olefins and respective products.

Feed	Operation	Catalyst	Products
Propylene	Dimerization	Aluminum alkyls	2-Methylpentene
Propylene	Dimerization	Basic metals	4-Methyl-1-pentene
Propylene	Trimers and tetramers	Phosphoric acid over a support	i-Nonenes and i-dodecenes
Propylene + n-butenes	Dimerization	Cobalt catalyst	i-Heptenes
i-Butene	Dimerization	Phosphoric acid over a support or ion exchange or H_2SO_4	Diisobutene

Butadiene

Butadiene is the supreme principal di-olefin in terms of use as a monomer in polymerization, and copolymerization in the industrial preparation of many types of rubber and thermoplastics. There is a lot of this compound for an appropriate price; hence it is a candidate for use as an intermediate in a number of industrial applications.

Historically, butadiene has been prepared from acetylene, in four steps: first conversion into acetaldehyde, then the latter is reacted through Aldol condensation to acetaldol, which is subjected to hydrogenation, and then the water is abstracted to give the final product required:

$$2\ CH_3CHO \xrightarrow{OH^{\ominus}} CH_3\text{-}\underset{\underset{\textstyle OH}{|}}{C}HCH_2CHO$$

$$\xrightarrow{+\ H_2} CH_3\text{-}\underset{\underset{\textstyle OH}{|}}{C}HCH_2CH_2OH \xrightarrow{-2\ H_2O} CH_2{=}CHCH{=}CH_2$$

129

The former socialist camp countries developed a wonderful method to produce butadiene in one step from ethanol. This method is still in place in Poland and Brazil. In this method, hydrogen is abstracted from ethanol, then dehydrated and dimerized in a single step, at 380°C, over a magnesium oxide catalyst on silica. This process is called Lepedeo's operation:

$$2\ CH_3CH_2OH \xrightarrow{\text{cat. 380 C}} CH_2=CH\text{-}CH=CH_2 + 2\ H_2O + H_2$$

Another method that starts from acetylene is its reaction with formaldehyde, followed by the hydrogenation of the product to give butane-1,4-diol, as an intermediate that is dehydrated to give butadiene:

$$HC\equiv CH + 2CH_2O \xrightarrow{+\ H_2} HOCH_2(CH_2)_2CH_2OH \xrightarrow{-2H_2O} CH_2=CH\text{-}CH=CH_2$$

Finally, butadiene is prepared through the removal of hydrogen by the oxidation of n-butanes and butenes. These are separated from C_4 fraction in the products of steam cracking.

Table (5-3) displays the content of butadiene (weight %) of the steam cracking of a number of feedstocks.

Table 5-3: The content of butadiene (weight %) in the steam cracking of a number of feedstocks.

Feed	Butadiene, wt.%
Ethane	1-2
Propane	4-7
n-Butane	7-11
Naphtha	12-15

Butadiene is chemically separated from C_4 fraction by formation of a complex of ammonium copper acetate $(Cu(NH_3)_2)OAc$. The complex disintegrates because the reaction is reversible. This method is used when the available quantities of butadiene are small. Butadiene is physically separated by extractive distillation through the addition of one of the following solvents that affect the relative volatility of the components of the mixture to be separated:

CH_3COCH_3 $CH_3CON(CH_3)_2$

CH_3CN $HCON(CH_3)_2$

Butadiene is synthesized from butane through Catadiene Process or from a mixture of butane and butenes as follows:

$$CH_3CH_2CH_2CH_3 + CH_2=CHCH_2CH_3 + CH_3CH=CHCH_3$$
$$\longrightarrow 3CH_2=CHCH=CH_3$$

The owner of this technology is ABB Lummus Company upon which it has established more than twenty factories during the last quarter of the twentieth century.

The significant uses of butadiene are in polymerization, whether homo- or copolymers. It gives the highly essential rubber, thermoplastic polymers and drying oils. The rubber that depends on butadiene can be produced in a wide spectrum of properties according to the co-monomer used. The most prominent of which is polystyrene-butadiene rubber used for production of small service car tires. Table (5-4) shows the international trend towards the production of synthetic rubber in the world (weight %). Figure (5-2) illustrates the main petrochemical derivatives of butadiene.

Butadiene is used in the synthesis of some intermediate chemical products through reactions such as addition, sulfonation and selective hydrogenation. Table (5-5) shows the most important uses of butadiene in 2011 in the entire world (ratio %).

Table 5-4: The international trend towards the production of synthetic rubber in the World (Wt.%).

Product	Production, wt. %
Styrene-butadiene rubber	36
Polybutadiene rubber	18
Chloroprene rubber	3
Olefin rubber (ethylene, propene, butadiene..)	8
Butyl rubber (with butadiene)	7
Isoprene rubber (with butadiene)	3
Nitryl rubber (with butadiene)	13
Other	12
Total	**100**

Table 5-5: The global butadiene consumption by application, 2011 (ratio %).

Product	Consumption %
Styrene-butadiene rubber (SBR)	32
Butadiene rubber	27
Hexamethylenediamine (HMDA)	6
Polyacrylonitrile-styrene-butadiene	9
Styrene-butadiene latex	10
Nitrile-butadiene rubber (NBR)	5
Other	11
Total	**100**

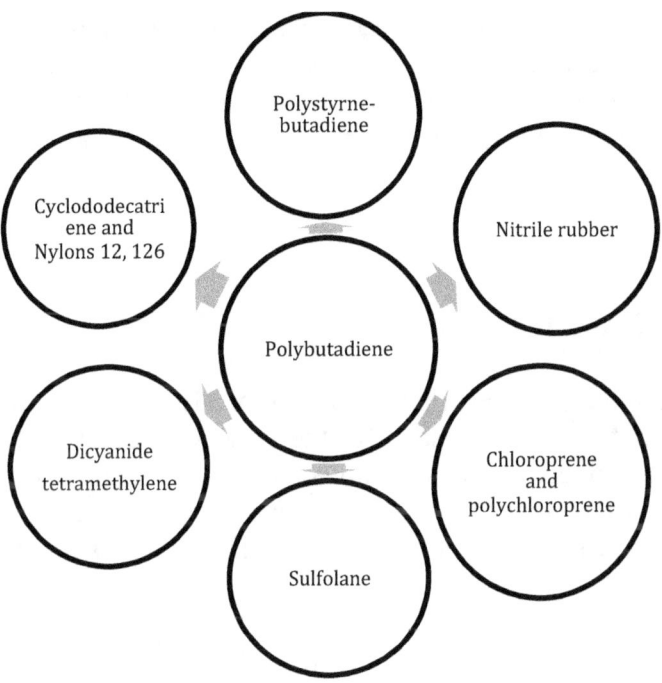

Figure 5-2: The main petrochemical derivatives of butadiene.

Isoprene

Isoprene is synthesized from isopentane or a mixture of isopentane and isopentene as in the following equation:

In one step, and from one feed or a combination of feeds, this wonderful selective catalysis process proceeds on a fixed bed, to produce the di-olefin. Catadiene catalysts are solid cylindrical balls made of a basic metal that give product yields ranging between 60 and 65% of the feed.

Butan-1,4-diol

The production of many industrial chemicals relied on the cheap acetylene before the flourishing of the petrochemical industry in the last century. Acetylene is substituted recently by ethylene in many applications due to its low price, abundance, ease of transport in pipelines, and safety, although ethylene is much less reactive than acetylene.

Calcium carbide used in the preparation of acetylene is still manufactured in many countries with cheap raw materials and abundant energy especially the free energy waterfalls. Acetylene is still manufactured in small quantities, where there are no alternatives to replace, such as in butanediol industry, which is at stake. Carbon black is manufactured from acetylene. It is used in the manufacture of batteries, an additive in the manufacture of rubber and plastics in general and in particular manufacture of tires. Calcium carbide is hydrated in large quantities of water for the preparation of acetylene. It is an easy method except that enormous quantities of water are wasted unless water is added in equivalent quantities.

The purification of acetylene is performed in three steps before use:

- Firstly, washing with sulfuric acid,
- Secondly oxidation by chlorohydrin or sulfuric acid,
- And finally washing with caustic soda and chlorohydrin (HOCl), or chlorine water, which must be chlorine free, so as not to react directly with the acetylene.

Acetylene can be obtained through the cracking of hydrocarbons. This process is endothermic. It requires a great deal of energy and a short time of reaction. The small quantities of acetylene formed in the cracking processes are washed by solvents.

Perhaps, the production of vinyls, esters, high alcohols, and the production of butan-1,4-diol in a smooth economic way, are the greatest significant uses of acetylene in the twentieth century, as in the following equations:

$$CH\equiv\!\!\equiv CH \ + \ 2\,HCHO \longrightarrow \ HOCH_2C\equiv\!\!\equiv CCH_2OH$$

$$HOCH_2C\equiv\!\!\equiv CCH_2OH \ \xrightarrow{\ 2\,H_2\ } \ HOCH_2CH_2CH_2CH_2OH$$

Table (5-6) summarizes the utmost principal uses of butane-1,4-diol.

Using associated technologics of oxidation, with fluid bed and hydrogenation of the fixed bed, n-butane can be converted to butanediol or a mixture of it with tetrahydrofuran, and/or gamma butyrolactone, as outlined in the following to the following equations:

$$CH_3CH_2CH_2CH_3 \longrightarrow HOCH_2CH_2CH_2CH_2OH$$

$$CH_3CH_2CH_2CH_3 \longrightarrow HOCH_2CH_2CH_2CH_2OH \ + \ \text{(tetrahydrofuran)}$$

$$CH_3CH_2CH_2CH_3 \longrightarrow HOCH_2CH_2CH_2CH_2OH \ + \ \text{(gamma butyrolactone)}$$

$$CH_3CH_2CH_2CH_3 \longrightarrow HOCH_2CH_2CH_2CH_2OH \ + \ \text{(tetrahydrofuran)} \ + \ \text{(gamma butyrolactone)}$$

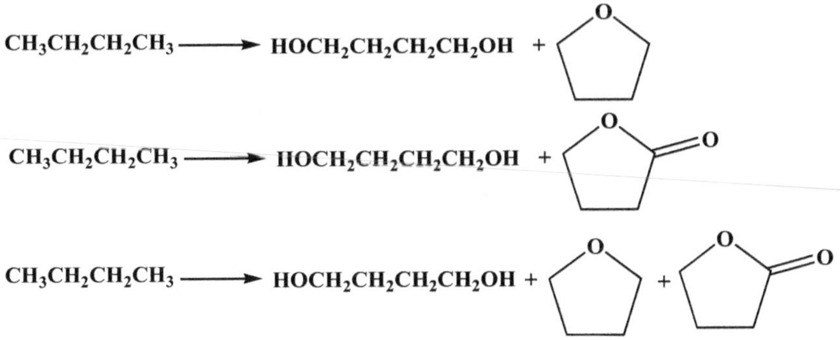

Table 5-6: The most important uses of butane-1,4-diol.

Product	Consumption%
Tetrahydrofuran	36
Vinyl chemicals (esters, alcohols, ketones)	18
Polybutene terephthalate	32
Polyurethanes	12
Other materials	2
Total	**100**

When n-butane is introduced with air into a catalyzed fluid bed reactor, it is oxidized to the maleic anhydride. This reactor gives a homogeneous temperature, for the catalyst to perform ideally and perfectly as shown in the following equation:

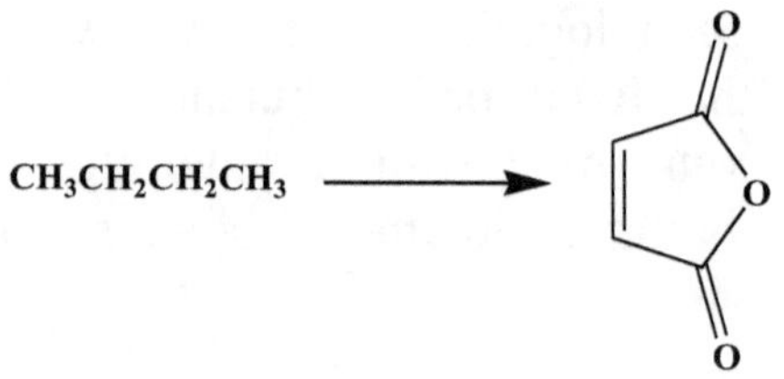

$$CH_3CH_2CH_2CH_3 \longrightarrow$$

In the water separator maleic anhydride is converted to maleic acid as follows:

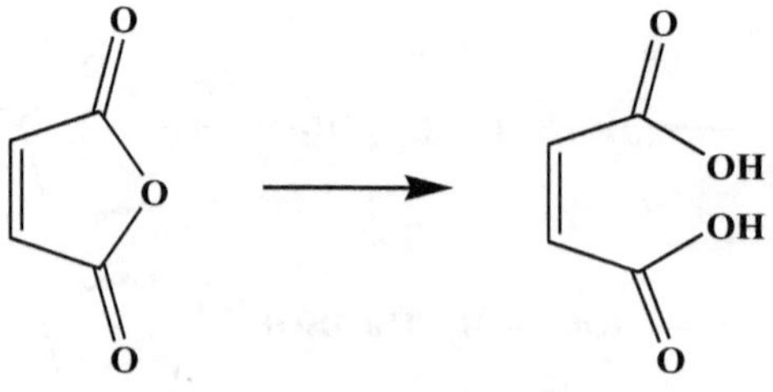

Maleic anhydride transforms entirely to maleic acid, therefore this method can be used for the production of maleic anhydride in the presence of the appropriate purification equipment. The gases

136

exhausting from the washing tower go to the incinerator to be disposed of safely.

The maleic acid product is sent to the hydrogenation catalytic reactor on the fixed bed. The yield is more than 94% of butanediol in a controlled hydrogen reactor, efficient extraction and purification system.

Tetrahydrofuran and butyrolactone can be obtained in acceptable and comparable ratios and yields. The following specific features characterize the process:

1) There is no continuous liquid waste for treatment; even the water formed during the purification process of the products is recycled to the water scrubber of maleic anhydride.

2) The catalysts do not need any treatments before nor after the operation. The process does not need addition of any other chemicals.

This technology is owned by the British Petroleum Company, which is well known for catalytic oxidation on the waterbed and the German Lurgi Oil-Gas Chemie GmbH with sound international experience in hydrogenation.

1-Butene

The French Petroleum Institute (IFP) has developed, in collaboration with the Foundation of Saudi Arabia Basic Industries Company (SABIC), the production process of high purity 1-butene, suitable for copolymers involved in polyethylene linear low density (LLDPE), through the process of dimerization of ethylene.

Alpha-butol Process is characterized by the ease of operation, high return, low operating conditions of pressure and temperature, liquid phase operation and concluded by using cheap carbon steel fittings.

The advantage of this method compared to similar operations is the continuous high-quality highly pure product, and reliable source of feed characterized by low capital cost, high yield, and ease of operation.

In the alpha-butol process, high purity ethylene is dimerized in a liquid phase reactor in a highly active and selective catalysis system. The liquid product is separated from the catalyst, and the liquid is distilled to recycle the unused ethylene, to remove and retrieve the volatile hydrocarbons. The purity of the final product exceeds 99.5%.

The owner of this technology is the French Petroleum Institute (IFP). Tens of units have been established that produce several hundred million tons per year.

n- and i-Butyraldehyde

The synthesis of n- and i-butyraldehyde is accomplished from propylene and synthesis gas, over a rhodium catalyst, in the IPOXO process under low pressure, according to the following equation:

$$CH_3CH=CH_2 + CO \xrightarrow{+H_2} CH_3CH_2CH_2CHO + CH_3CH(CH_3)CHO$$

In this process, propylene is reacted with synthesis gas of 1:1 ($CO:H_2$) ratio, under low pressure (greater than twenty kg/sq. cm.), in the presence of a catalyst of rhodium, complexed with one ligand, to give iso and n-butyraldehyde in a ratio of 1:10.

Other products can be added to this technology such as n- and i-butyl alcohol, Aldolization, and hydrogenation to 2-ethyl hexanol:

$$CH_3CH_2CH_2CHO + H_2 \longrightarrow CH_3CH_2CH_2CH_2OH$$

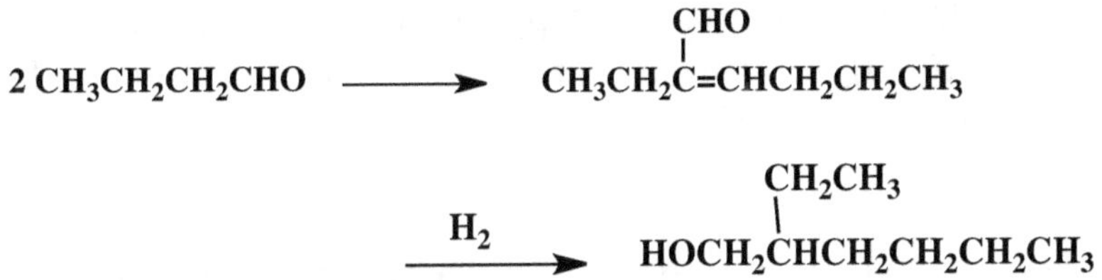

$$2\ CH_3CH_2CH_2CHO \longrightarrow \overset{\displaystyle CHO}{\underset{|}{CH_3CH_2C}}=CHCH_2CH_2CH_3$$

$$\xrightarrow{\ H_2\ } HOCH_2\overset{\displaystyle CH_2CH_3}{\underset{|}{C}}HCH_2CH_2CH_2CH_3$$

Butyl alcohols

There are four butyl alcohols:

n-Butanol $CH_3CH_2CH_2CH_2OH$

i-Butanol $CH_3(CH_3)CH_2CH_2OH$

2-Butanol $CH_3CH_2CH(OH)CH_3$

2-Methyl-2-propanol $(CH_3)_3COH$

n-Butanol is synthesized in four ways:

1) Through the hydroformylation of propylene:

$$CH_3CH=CH_2 + CO \xrightarrow{+ H_2} CH_3CH_2CH_2CHO + CH_3CH(CH_3)CHO$$

$$\xrightarrow{+ H_2} CH_3CH_2CH_2CH_2OH + CH_3CH(CH_3)CH_2OH$$

2) Through Aldol condensation of acetaldehyde:

$$2\ CH_3CHO \longrightarrow CH_3CH=CHCHO \xrightarrow{+ H_2} CH_3CH_2CH_2CH_2OH$$

3) Through sugar or starch fermentation.

4) The method of Reppe, through the reaction of propylene with carbon monoxide and water in the presence of a modified catalyst of iron pentacarbonyl. In this Reppe hydrocarboxylation CO and H_2 (from $Fe(CO)_5$ and H_2O) are transferred to propene from an intermediate Fe-CO-H complex:

$$2\ CH_2{=}CHCH_3 + 4\ CO + 2\ H_2O \xrightarrow{-2\ CO_2} HOCH_2CH_2CH_2CH_3 + HOCH_2\overset{\overset{\displaystyle CH_3}{|}}{C}HCH_3$$

Secondary butanol (2-butanol) is synthesized through the direct hydration of n-butene, in the presence of sulfuric acid, through the formation of alkyl sulfates, followed by hydration:

$$CH_3CH_2CH{=}CH_2 + CH_3CH{=}CH_2CH_3 \xrightarrow{+H_2O + H_2SO_4} CH_3CH_2\overset{\overset{\displaystyle OH}{|}}{C}HCH_3$$

The same method can be used in the synthesis of t-butanol from isobutane, but in the case n-butenes a concentration of 75-80% sulfuric acid is used, while in the case of i-butene the reaction proceeds in a concentration of 50-60%:

$$CH_2{=}\overset{\overset{\displaystyle CH_3}{|}}{\underset{\underset{\displaystyle CH_3}{|}}{C}} \xrightarrow{H_2O + H_2SO_4} H_3C-\overset{\overset{\displaystyle CH_3}{|}}{\underset{\underset{\displaystyle CH_3}{|}}{C}}-OH$$

Secondary and tertiary butane alcohols are used as solvents in the manufacture of paint, anti-freeze liquids, additives to motor gasoline, to prevent detonation and improve quality, in addition to some organic intermediates, such as the removal of hydrogen from 2-butanol, to be transformed into methyl ethyl ketone, and the introduction of t-butyl group on the benzene ring in the preparation of antioxidants.

140

Vinyl ethers

The easiest way to prepare vinyl ethers is through Reppe's method by reacting an alcohol with acetylene, in the presence of potassium alcohols as catalysts:

$$\text{ROH} + \text{HC}\equiv\text{CH} \longrightarrow \text{ROCH}=\text{CH}_2$$

Polyvinyl ethers are used as polymers or copolymers as additives in the manufacture of paint, stickers, and artificial leather.

Chapter Six
Petrochemicals from Aromatics

PETROCHEMICALS FROM AROMATICS 144
Catalytic reforming..145
Continuous catalytic reforming 147
Cyclar process...148
Pyrolysis gasoline...149
SEPARATION PROCESSES 150
Solvent extraction..151
Azeotropic distillation...152
Fractional distillation..152
Extractive distillation 152
Crystallization 154
Solvent-solvent extraction 154
PETROCHEMICALS FROM BENZENE...............................155
Bisphenol A 159
Cumene...161
Linear alkyl benzene 163
Cyclohexane...164
Phenol..165
Nitrobenzene 169
Maleic anhydride 170
PETROCHEMICALS FROM TOLUENE 173
Benzoic Acid..174
Trinitrotoluene..175
ε-caprolactam..175
PETROCHEMICALS FROM XYLENES.............................178
Isomerization of Xylenes..178
Phthalic Anhydride...179
Terephthalic Acid 181
Dimethyl Terephthalate...187

Petrochemicals from Aromatics

The term "Aromatics" represents the total benzene, toluene and xylene compounds, which also include ethyl benzene. The total xylenes and ethyl benzene are also called C_8 fraction.

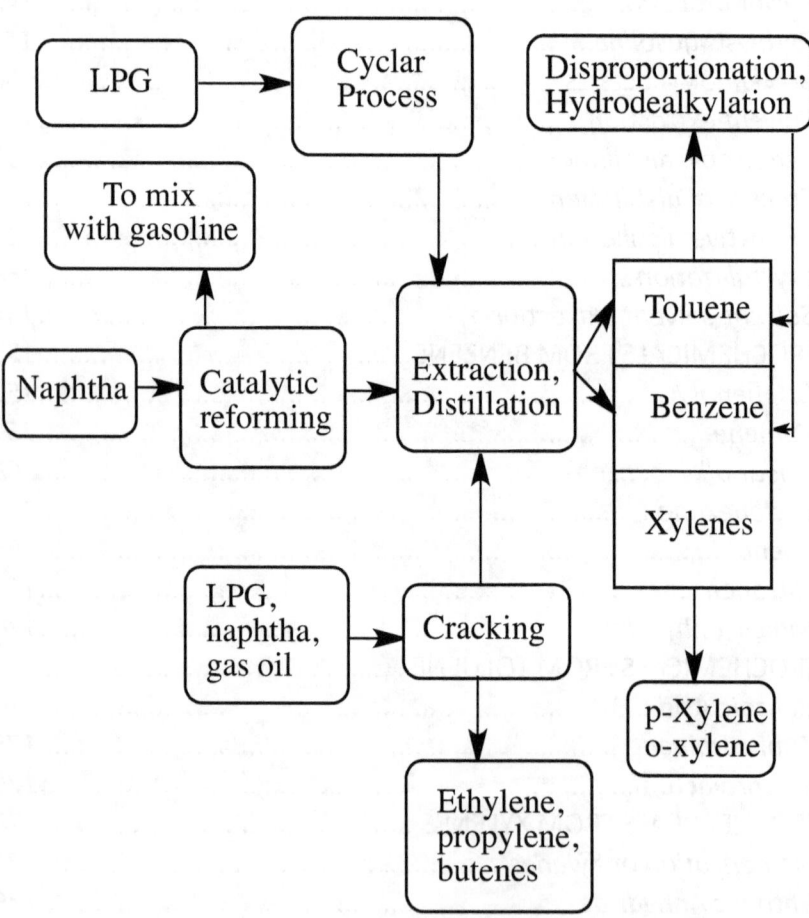

Figure 6-1: The syntheses of aromatics from oil derivatives.

Aromatics are considered one of the greatest significant components of the petrochemical industry, since three primary petrochemicals belong to them: benzene, toluene and xylenes. The world used to derive its few needed aromatics from coal tar until the end of the Second World War, where it started to draw aromatics from oil fractions, when that became achievable scientifically, technically and economically applicable.

There are three modern petrochemical sources of aromatics:

1. Catalytic reforming including continuous catalytic reforming.

2. Pyrolysis gasoline and its source is the pyrolysis of naphtha and LPG.

3. LPG conversion process of aromatics through the famous UOP technology in which butane and propane are converted into aromatics; this process is also called Cyclar process.

Figure 6-1 summarizes the most important and prominent industrial pathways in the aromatics industry from oil derivatives.

Catalytic reforming

Catalytic reforming is practiced in almost all refineries in the World, as a basic technique prevailed whereby naphtha, which contains alkanes and cyclic alkanes, is converted into aromatics, for the production of the appropriate specification of motor gasoline for the operation of automobiles, especially with regard to the octane number rating, and the smooth ignition in the internal combustion engines. It is well known that the aromatic compounds have high octane numbers and their presence evidently improves the quality of motor gasoline.

Catalytic reforming is used in practically all petrochemicals refineries in the World furthermore for the production of aromatic hydrocarbons. Naphtha contains mainly six to eight carbon compounds (C_6-C_8 fractions). These are selected for this reaction, and the appropriate conditions are applied to provide large yields of the three important primary petrochemicals: benzene, toluene and the xylenes.

Some of the greatest significant chemical processes that take place in the catalytic reforming are summarized below:

1) Isomerization

2) Dehydrogenation

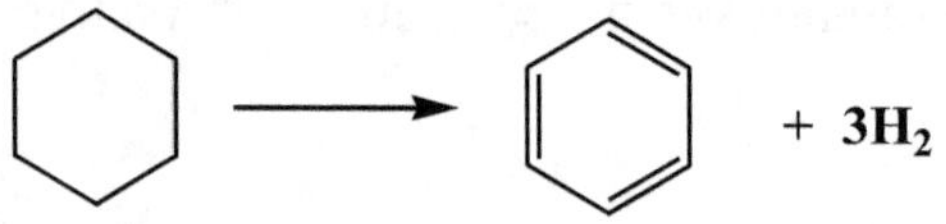

2) Cyclization by dehydrogenation

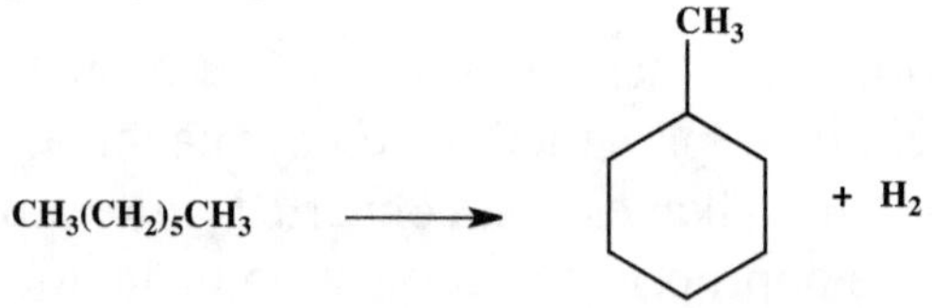

4) Dealkylation

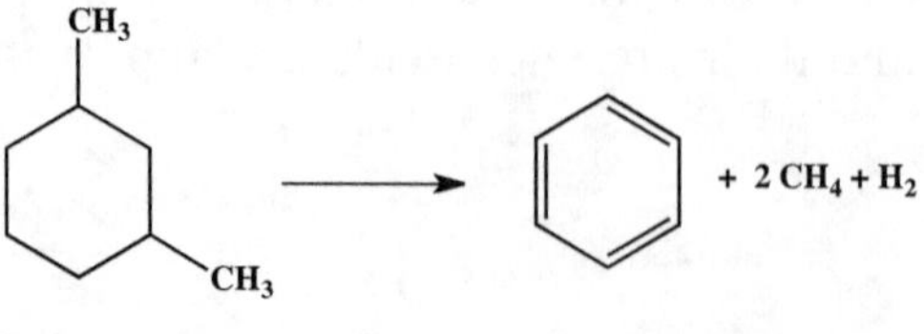

146

5) The presence of hydrogen, produced in the processes, leads to the hydrogenation of the olefins present, the hydrocracking of paraffins, and the removal of sulfur, as illustrated in the following equations:

$$CH_3(CH_2)_4CH=CH_2 \ + \ H_2 \longrightarrow CH_3(CH_2)_5CH_3$$

$$C_6H_{14} \ + \ H_2 \longrightarrow 2CH_3H_8$$

$$CH_3CH_2CH_2SCH_2CH_2CH_3 \ + \ H_2 \longrightarrow 2\,CH_3CH_2CH_3 \ + \ H_2S$$

Continuous catalytic reforming

Benzene, toluene and xylenes are produced in large revenues through the continuous catalytic reforming process. This process is associated with another important task that is the continuous regeneration of the catalyst. The products are pumped to the aromatics extraction unit for separation. The xylenes fraction is manipulated to acquire pure para and ortho xylenes through fractional distillation, followed by hydrogenation to remove olefins and diolefins.

A reactor with a mobile layer and a catalyst regeneration system linked with a strong easy to move catalyst characterize this process. The plant is fully automatic and requires only little supervision and control.

Table 6-1 summarizes the final yields of the continuous catalytic reforming process (weight%).

The aromatics are separated from each other through extractive distillation in a modern technique owned by the French Petroleum Institute (IFP). It established more than six major plants in the World.

Table 6-1: The yields of the continuous catalytic reforming process (weight%).

Feed (Naphtha 80 – 150 °C)		Products	
Paraffins	57	Hydrogen	4.1
Naphthenes	37	C5+	87
Aromatics	6	Benzene	8.5
		Toluene	26.3
		Xylenes	26.1
		Total aromatics	74.3

Cyclar process

British Petroleum (BP) and Universal Oil Prospectus (UOP) have developed the Cyclar process. This wonderful technique converts propane and butane (liquefied petroleum gas, LPG) into aromatics; the utmost significant primary petrochemicals: benzene, toluene and xylenes:

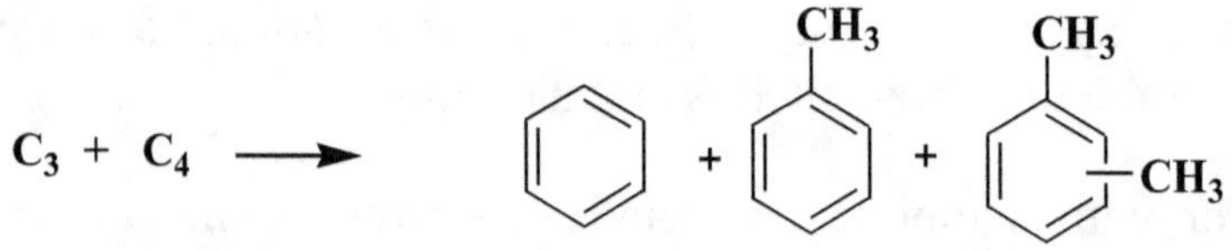

The process consists of three sections: the reactor, the continuous catalyst regeneration unit, and the product extraction section. Table (6-2) summarizes the yields in Cyclar process as feed (weight %).

There is a working factory of this kind in the Arab World in Yanbu, Saudi Arabia with a production capacity of six to forty thousand barrels per day. This technology is owned by UOP.

Table 6-2: The yields in the Cyclar process according to the feed (weight%).

Feed: propane only	61
Different butanes only	66

Products, % from feed	
Hydrogen	7
Benzene	27
Toluene	43
Aromatics C_8	22
Aromatics C_9	8

Pyrolysis gasoline

Another basic method for the production of aromatics from oil is through the pyrolysis gasoline. Pyrolysis gasoline is an oil fraction in the motor gasoline boiling range obtained as a by-product from the cracking of naphtha or gas oil, or even cracking ethane for ethylene and some other olefins such as propylene and butenes. The aromatics are extracted from pyrolysis gasoline using the typical known techniques. It is to be noted that a number of olefins and di-olefins must be disposed of though hydrogenation before the extraction of the aromatics by polar solvents.

Pyrolysis gasoline is characterized by the presence of a high percentage of aromatics when compared with catalytic reforming, a higher-priced, and much needed, in the production of many petrochemical products, as will be detailed later. Table (6-3) shows the distribution of the ideal components of the pyrolysis gasoline, catalytic reforming and Cyclar process yields.

Table 6-3: The typical distribution of the components of pyrolysis gasoline, catalytic reforming and Cyclar Process.

Components	Pyrolysis gasoline	Catalytic Reforming	Cyclar Process
Benzene	39	3	27
Toluene	20	13	43
Xylenes	5	18	22
Ethylbenzene	2	5	-
Large aromatics	3	16	8
Non-aromatics	31	45	-
Total	**100**	**100**	**100**

Separation processes

Purification of aromatics takes place in two steps:

The first is the separation of the non-aromatics from aromatics.

The second step is the separation of the aromatic compounds themselves from each other.

There are several alternatives and different techniques for the separation. The separation methods depend on the ratio of aromatics in the mixture and their nature as well as perhaps the appropriate use of a number of different ways to separate some compounds such as separation of the elements of the aromatic fraction of C_8 in a series of distillation and crystallization, as seen in Table (6-4).

Table 6-4: Some natural properties of C_8 fraction aromatics.

Compound	Melting point °C	Boiling point °C
Ethylbenzene	-95.0	136.2
o-Xylene	-25.2	144.4
m-Xylene	-47.9	139.1
p-Xylene	13.3	138.4

Table 6-5: The main aromatics extraction operations.

Process	Separation	Operation requirements
Isotropic distillation	Separate BTX from Pyrolysis gasoline	>90% Aromatics
Extractive distillation	Separate BTX from Pyrolysis gasoline	65-90% Aromatics
Solvent-solvent extraction	Separate BTX from Pyrolysis gasoline	20-65% Aromatics
Crystallization by freezing	Separate p-xylene from o- and m- xylene	Separate o-xylene and ethyl benzene from C_8 fraction
Adsorption on solid	Separate p-xylene from C_8 fraction	Continuous selective separation and reversible

Thus, ethyl benzene could be separated by fractional distillation at 136.4°C and then p-xylene is separated by fractional cooling of the mixture at -60°C. o-Xylene could be separated at 144.4°C. Thus, we are left with only m-xylene. The most significant aromatics separation operations are summarized in Table (6-5).

Solvent extraction

The solvent extraction is the first step in the separation of a high concentration of aromatics from non-aromatics. This depends on the selective nature of the selected solvent used to separate aromatics from the non-aromatics in two layers that could be easily separated in an appropriate temperature. The solvent added increases the difference between the boiling points of the aromatics and the non-aromatics and increases the volatility of the aromatics in particular. The solvent used has to be non-corrosive, non-reactive, and thermally stable in use. The aromatics are separated from the solvent first and the solvent remains at the bottom. Table 6-6 summarizes some of the most commonly used solvents in the separation of aromatics.

Azeotropic distillation

The idea of separating the aromatics from the non-aromatics by azeotropic distillation relies on the use of a highly polar solvent, such as acetone or methanol to increase the volatility of the non-aromatic (an azeotrope is formed) and distills out with it, leaving the aromatics at the bottom of the distillation column. This method is used when the aromatics constitute a high proportion of the mixture and the small quantity of the non-aromatics is distilled. Azeotropic distillation is used extensively in the separation of 10% of the non-aromatics from aromatics (90%) in the pyrolysis gasoline.

Fractional distillation

Aromatics can be separated from each other by fractional distillation. Theoretically, any two compounds can be separated from each other provided that they have two different boiling points depending on the length of distillation column and the number of dishes in the column.

Extractive distillation

In the extractive distillation, the extraction is applied in the presence of a solvent that is less volatile than the entire components of the mixture to be separated. The solvent is selected so that it makes the component to be separated relatively more volatile than the other components of the mixture. Naturally, the proposed solvent must experience some advantages such as non-reactive with the mixture components and does not corrode the used equipment in addition to the thermal stability, efficiency and its boiling point must be higher than all the mixture components after playing its role.

Table (6-7) summarizes examples of three solvents used in the extractive distillation of toluene, benzene and styrene.

Table 6-6: Some solvents used in the extraction of aromatics.

Solvent	Composition	Bp. °C
Furfural	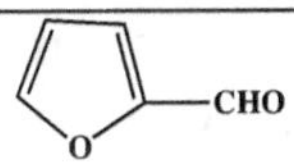 (furan ring)—CHO	162
Sulfur dioxide	SO_2	-
Diethylene glycol	$HOCH_2CH_2OCH_2CH_2OH$	345
Tetra ethylene glycol	$HO(CH_2CH_2O)_3CH_2CH_2OH$	315
Sulfolane	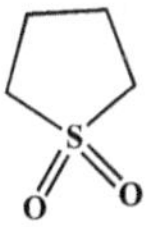 (ring structure with S, O, O)	285
N-Methylpyrrolidone	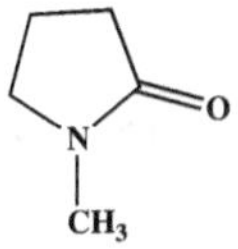 (ring structure with N, O, CH₃)	302
Dimethylsulfoxide	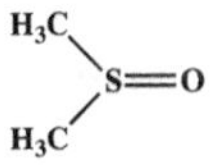 H_3C, H_3C S=O	189
N-Formylmorpholine	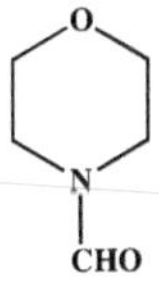 (ring structure with O, N, CHO)	-

153

Table 6-7: Examples of extractive distillation.

Solvent	To separate	From
Phenol	Toluene	A small fraction of catalytic reformate
Dimethylformamide	Benzene	C_6 fraction of pyrolysis gasoline
Dimethylacetamide	Styrene	C_8 fraction of pyrolysis gasoline

Crystallization

Crystallization is used in the separation of the components of a mixture with a large difference between their melting points. That is particularly essential in the separation of p-xylene from the other xylene isomers.

Solvent-solvent extraction

In this technique, a mixture of polar solvents is used leading to the formation of two immiscible layers that can be separated and distilled. It is to be noted that there is only one layer in the case of extractive distillation while there are two layers in solvent-solvent extraction and that is the main difference between them.

PETROCHEMICALS FROM BENZENE

Benzene is one of the primary petrochemicals. It is considered an imperative intermediate in organic aromatic chemicals, and has multiple uses and applications in the chemical industry. About 80% of the benzene is consumed in the manufacture of ethyl benzene, cumene and cylohexane and the remainder is used in the industrial preparation of nitrobenzene, maleic anhydride and dodecyl benzene. Figure 6-2 summarizes the global benzene capacity by process in 2010.

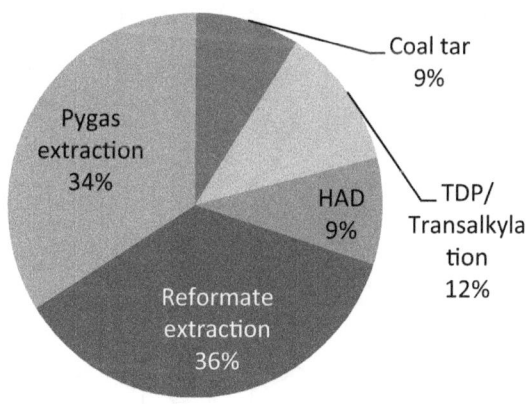

Figure 6-2: The global benzene capacity by process, 2010.

Figure 6-3 summarizes the most important derivatives and the modern sources of benzene. We have seen in the previous chapters the manufacturing of styrene from ethyl benzene and polystyrene from styrene, as well as phenol and acetone from cumene, as we also knew that dodecyl benzene is used in the manufacture of detergents.

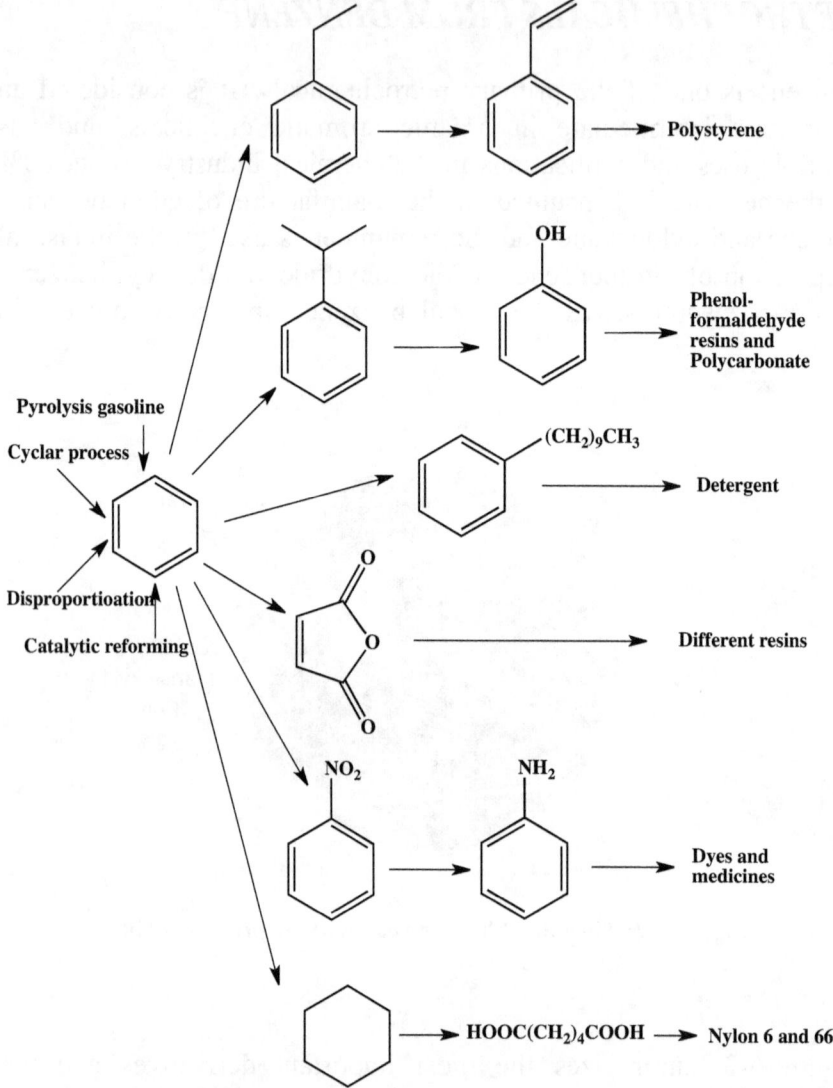

Figure 6-3: The most important industrial derivatives of benzene.

As stated previously, benzene, as well as o- and p-xylenes, produced in the catalytic reforming processes and pyrolysis gasoline, did not meet the enormous demand required in the petrochemical industry. At the same time, the production of toluene and m-xylene is much more than needed in the petrochemical industries. Although toluene and m-xylene are used as good solvents, as well as added to motor gasoline, to improve the quality, and raise the octane rating, they constitute great sources for the manufacture of benzene and p-xylene in particular, but through different techniques and methods.

In the toluene hydrodealkylation method, toluene is reacted thermally with hydrogen at 580-800°C, under the pressure of 30-100 atmospheres, in the presence of a catalyst of chromium, manganese and cobalt oxides for the production of benzene and methane:

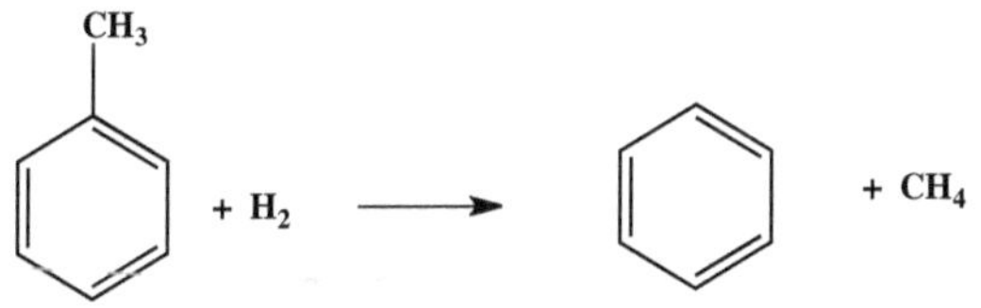

The other method to convert toluene to benzene is a dual exchange of toluene called toluene disproportionation. In this process, at 80-125°C, and under a pressure of 50-70 atmospheres, and in the presence of a Lewis acid catalyst, a toluene molecule provides its methyl group to another toluene molecule. The products are a benzene molecule and a xylene molecule according to the following equation:

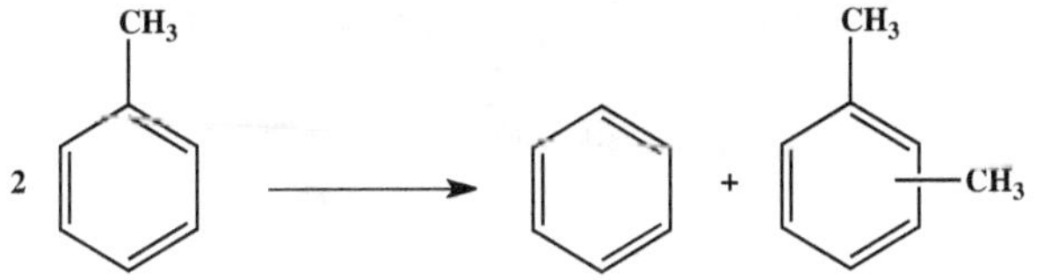

Dettol is the newly introduced method used in the production of benzene from toluene and heavier aromatics. The same method can be used to produce xylenes in addition to benzene. The yield of this process exceeds 99% of toluene or the heavy aromatics used in the reaction.

Table 6-8: The production yields of benzene and xylene by Dettol process.

Feed, wt%	Type of product	
	Xylene	Benzene
Non-aromatics	2.3	3.2
Benzene	11.3	
Toluene	0.7	47.3
Aromatics C_8+	49.5	0.3
C_9+ Aromatics	-	85.4
Products as wt.% of feed		
Benzene	75.7	36.9
Aromatics C_8	-	37.7

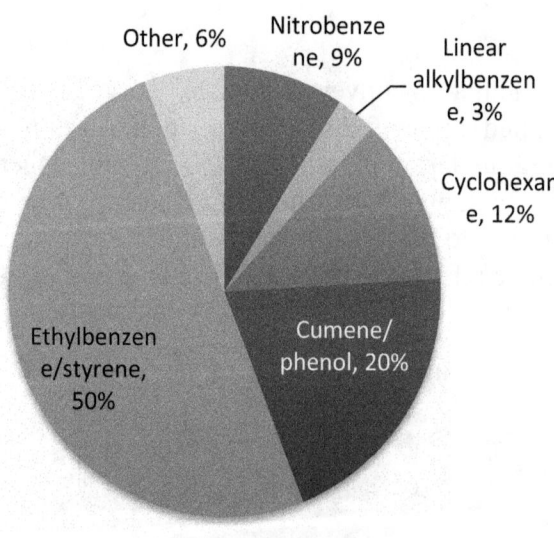

Figure 6-4: The World consumption of benzene, 2010.

Table (6-8) summarizes the production yields of benzene and xylene by Dettol process. Figure 6-4 represents the ideal distribution of the principal uses of benzene in the World in 2010 (ratio %).

ABB Lummus owns this Dettol technology and has established tens of factories with production ranges between twelve and one hundred million gallons per year.

Bisphenol A

Bisphenol A is used in the manufacture of polycarbonate and in resin industry. It is manufactured from phenol and acetone as shown in the following equation:

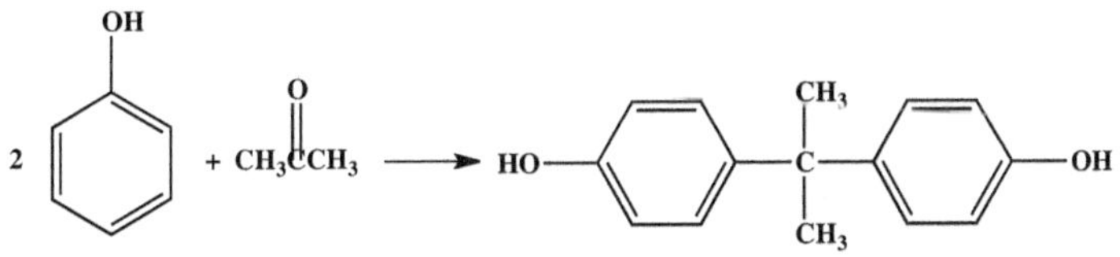

The catalyst has a long life and the ability to convert a maximum amount of acetone to bisphenol A. It reduces the by-products, and the volume of the catalyst. This technology is owned by Chiyoda Foundation, Japan. It has established many factories in Japan and Taiwan and other countries. The annual production varies between twenty-five and seventy thousand tons.

Bisphenol A finds wide use and applications in some polymers industry as shown in the following equations:

1) Polysulfone:

$$\text{HO}-C_6H_4-\underset{\underset{\text{CH}_3}{|}}{\overset{\overset{\text{CH}_3}{|}}{C}}-C_6H_4-\text{OH} \;+\; \text{Cl}-C_6H_4-\underset{\underset{O}{\parallel}}{\overset{\overset{O}{\parallel}}{S}}-C_6H_4-\text{Cl}$$

$$\longrightarrow \;\Big[\!-O-C_6H_4-\underset{\underset{\text{CH}_3}{|}}{\overset{\overset{\text{CH}_3}{|}}{C}}-C_6H_4-O-C_6H_4-\underset{\underset{O}{\parallel}}{\overset{\overset{O}{\parallel}}{S}}-C_6H_4-\!\Big]_n$$

3) Polycarbonate:

$$\text{HO}-C_6H_4-\underset{\underset{\text{CH}_3}{|}}{\overset{\overset{\text{CH}_3}{|}}{C}}-C_6H_4-\text{OH} \;+\; \text{Cl}-\underset{}{\overset{\overset{O}{\parallel}}{C}}-\text{Cl} \longrightarrow$$

$$\Big[\!-O-C_6H_4-\underset{\underset{\text{CH}_3}{|}}{\overset{\overset{\text{CH}_3}{|}}{C}}-C_6H_4-O-\overset{\overset{O}{\parallel}}{C}-\!\Big]_n$$

3) Epoxy resins basics:

$$\text{HO}-C_6H_4-\underset{\underset{\text{CH}_3}{|}}{\overset{\overset{\text{CH}_3}{|}}{C}}-C_6H_4-\text{OH} \;+\; \text{ClH}_2C-\underset{\underset{H}{|}}{\overset{\overset{O}{\diagdown}}{C}}\!-CH_2 \longrightarrow$$

$$H_2C\!\overset{O}{\diagup\!\diagdown}\!\underset{H}{C}-\overset{H_2}{C}-O-C_6H_4-\underset{\underset{\text{CH}_3}{|}}{\overset{\overset{\text{CH}_3}{|}}{C}}-C_6H_4-O-\overset{H_2}{C}-\underset{H}{C}\overset{O}{\diagdown\!\diagup}CH_2 \longrightarrow \longrightarrow$$

4) Polyaryls:

$$\text{HO}-C_6H_4-\underset{\underset{\text{CH}_3}{|}}{\overset{\overset{\text{CH}_3}{|}}{C}}-C_6H_4-\text{OH} \;+\; C_6H_4(\text{COOH})_2 \longrightarrow$$

$$\Big[\!-O-C_6H_4-\underset{\underset{\text{CH}_3}{|}}{\overset{\overset{\text{CH}_3}{|}}{C}}-C_6H_4-O-\overset{\overset{O}{\parallel}}{C}-C_6H_4-\overset{\overset{O}{\parallel}}{C}-\!\Big]_n$$

Cumene

Cumene is of paramount significance in the world of petrochemicals as the main feedstock of phenol. He latter is used in the production of a number of cross-linked polymers known as phenol-formaldehyde resins among a number of other principal final products and performance polymers.

The Hock Process is the most important method in the world to produce cumene. It consists of three steps:

1) The reaction of propylene with benzene for the production of cumene.

2) The oxidation of cumene.

3) The acidification of the hydroperoxide formed to give phenol and acetone products.

The production of cumene starts from the alkylation of benzene with propylene gas, as shown in the following equation:

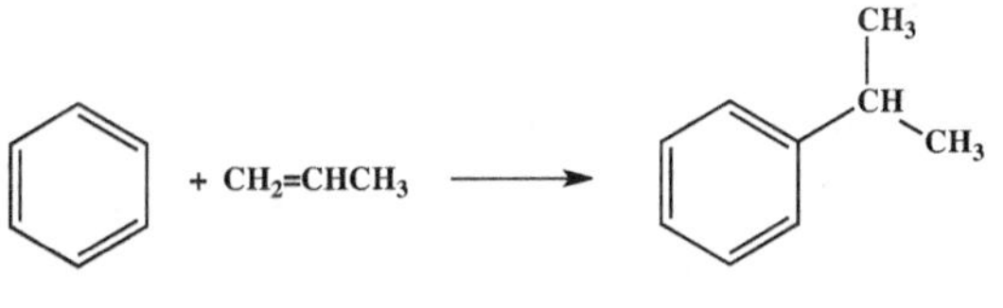

$$\text{C}_6\text{H}_6 + \text{CH}_2\text{=CHCH}_3 \longrightarrow \text{C}_6\text{H}_5\text{CH(CH}_3)_2$$

The most prominent international producers utilize various Zeolite catalysts in this conversion, which is characterized by catalytic practicability of recovery, and non-ability to corrosion that reduces the cost of construction, and permits the use of regular steel reactors.

One of the most important problems in alkylation is that the alkylated products (cumene) become able to add another alkyl group faster than benzene itself, and thus the reaction leads to poly-alkylated benzenes:

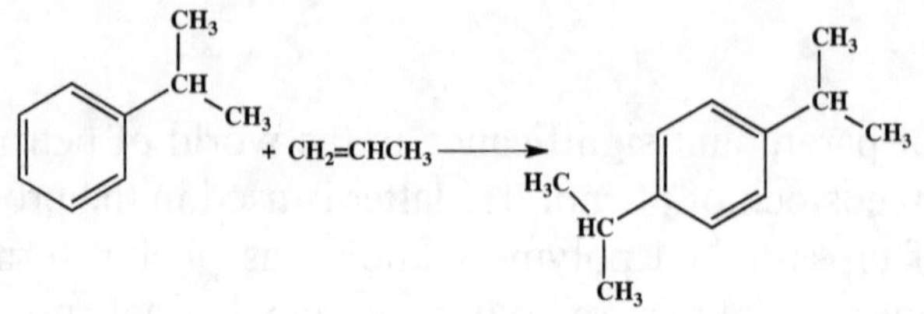

The global manufacturers are keen to convert these polyalkylated benzenes to cumene through trans-alkylation as illustrated in the following equation:

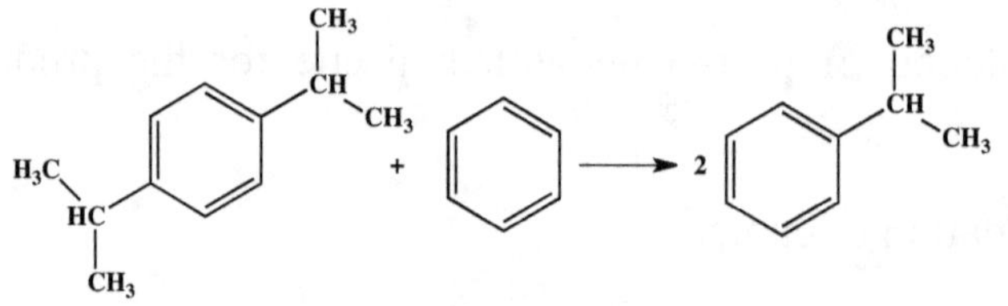

This is to get rid of the polyalkylated product in one hand, and increase the final product, cumene, in addition to reduction in impurities, and exclusion of the costs of separation and purification charge. The following equations show the rest of Hock process:

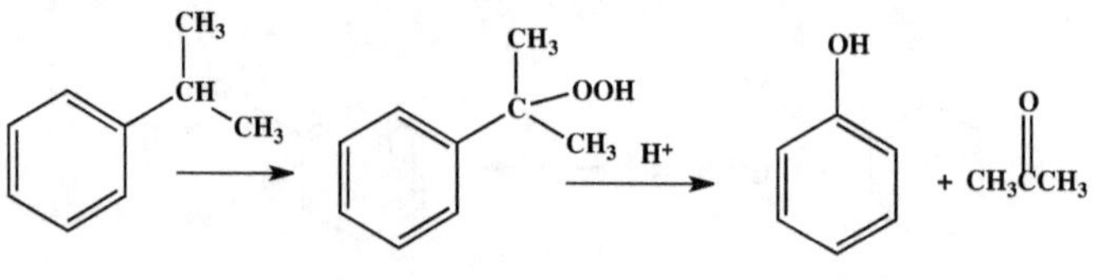

Hock process can be used in the production of the following compounds too:

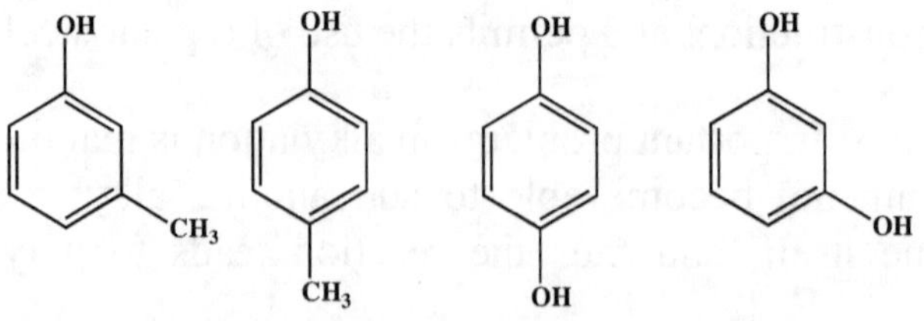

Linear alkyl benzene

Monoalkylated benzene with an alkyl group varying in length between 10 and 14 carbons is a suitable raw material for the production of linear alkyl benzene sulfonates, which have great significance in the detergent industry.

This industry has adopted the dodecyl benzene sulfonates, which is synthesized from the condensation of four propylene compounds in the presence of supported phosphoric acid as a catalyst, at 200-240°C, and under the pressure of 15-25 atmospheres. But this compound degrades biologically with difficulty, and thus polluting the environment. It is replaced by isoalkylbenzene sulfonate, which biodegrades easily, because of the absence of branches.

Normal olefins that contain 10-14 carbons may be obtained from normal paraffins using one of the following methods:

1) Thermal cracking of wax.

2) Chlorination and then removing hydrochloric acid.

3) Dehydrogenation.

4) Oligomerization of ethylene using Ziegler-Natta catalysts.

Benzene is alkylated with normal olefins in the liquid phase, usually at 40-70°C, in the presence of a Lewis acid catalyst such as HF, HBF_4 and $AlCl_3$. It could also be done in the gas phase in the presence of a Lewis acid catalyst. In the next step the alkyl benzene is sulfonated by sulfur trioxide, or oleum, then converted to the sodium salt:

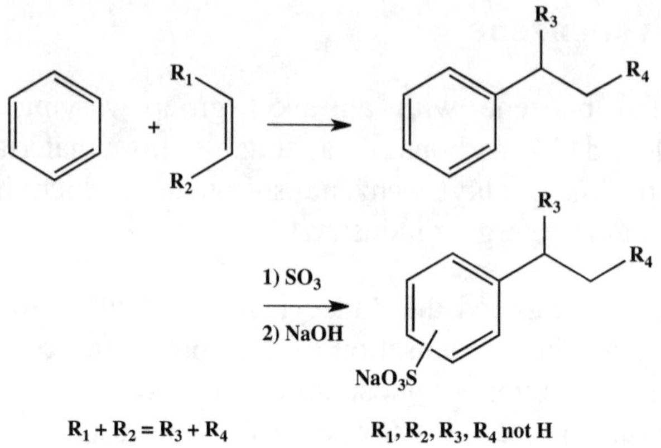

$R_1 + R_2 = R_3 + R_4$ R_1, R_2, R_3, R_4 not H

Cyclohexane

In the traditional petroleum industry, small quantities of cyclohexane are obtained from the oil fraction using the selective extraction technique and through the isomerization of methyl cyclopentane to cyclohexane as follows:

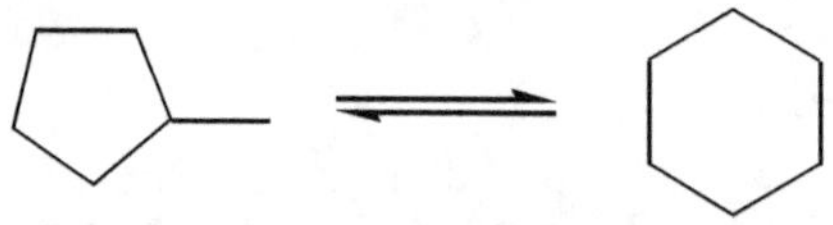

Recently, pure benzene is hydrogenated on a nickel catalyst at 170-230°C and under a pressure of 20-40 atmospheres, for the synthesis of cyclohexane. The latter is obtained by hydrogenation of benzene in the liquid phase according to the following equation:

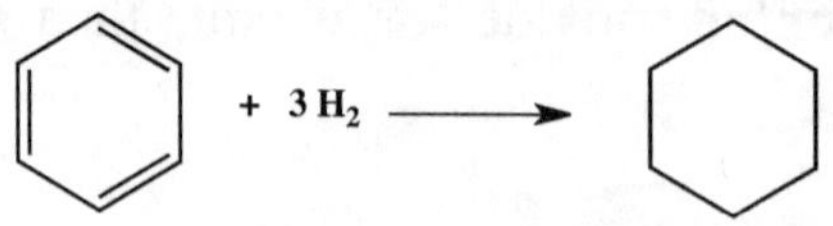

All the feed is converted in the reactor, at low temperature, using an insoluble catalyst with continuous injection. The high catalytic efficiency allows the use of a relatively low partial pressure of

164

hydrogen, leading to fewer side reactions like isomerization, and hydrocracking. The reaction temperature allows the evaporation of the cyclohexane product, and through a recycling pump around the heat exchanger a low partial pressure steam is produced; this allows the control of the reaction temperature. In the last reactor, which operates in the vapor phase, the remaining benzene is converted to cyclohexane by catalytic hydrogenation. This step reduces the benzene percentage in the cyclohexane product to a low level. Depending on the purity of the hydrogen used, an LPG separator is installed in the stabilization section, or a small separator to remove the produced light gases. This method produces 1,075 grams of cyclohexane from each kilogram of benzene.

This technology is owned by the French Petroleum Institute and was used to establish tens of factories for the production of cyclohexane.

Perhaps the most important use of cyclohexane is its oxidation to adipic acid in several steps. Adipic acid is used in the manufacture of nylon 66 through its reaction with hexamethylenetetraamine:

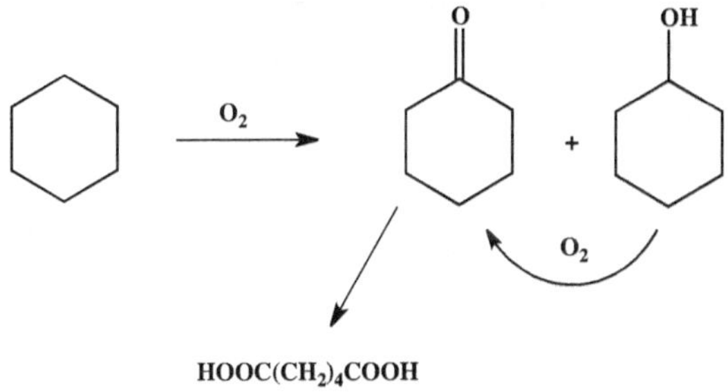

$$HOOC(CH_2)_4COOH$$

Phenol

We mentioned in chapter four of the petrochemicals from propylene the prevailing manufacturing method of cumene throughout the World, for the preparation of phenol and acetone. When no appropriate discharge is found for acetone, cyclohexene is used

165

instead to substitute for propylene in the reaction with benzene and then the same steps continue as in the case of cumene. The hexanone product formed, instead of acetone, is hydrogenated to cyclohexanol then dehydrated to cyclohexene and re-used again and again. The following equations show the pathway of this process:

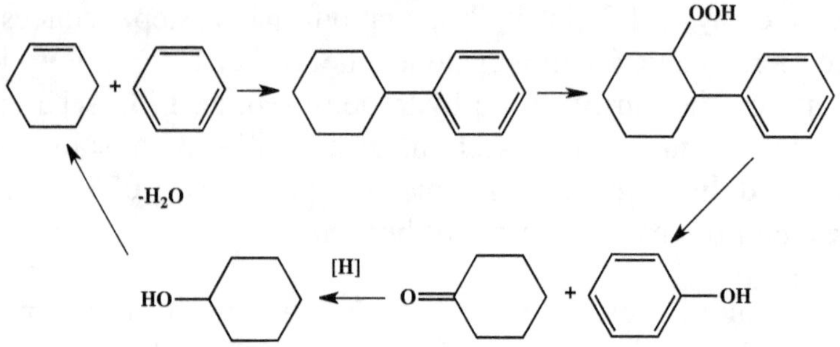

The other relatively recent method benefits from the cheap toluene which is oxidized to benzoic acid at 110°C, in the presence of a cobalt salts catalyst, under 2-3 atmospheres of pressure, and then the benzoic acid is oxidized to phenol at 250°C under atmospheric pressure, in the presence of copper salts as catalyst. In addition, carbon dioxide is produced, which is used in several applications:

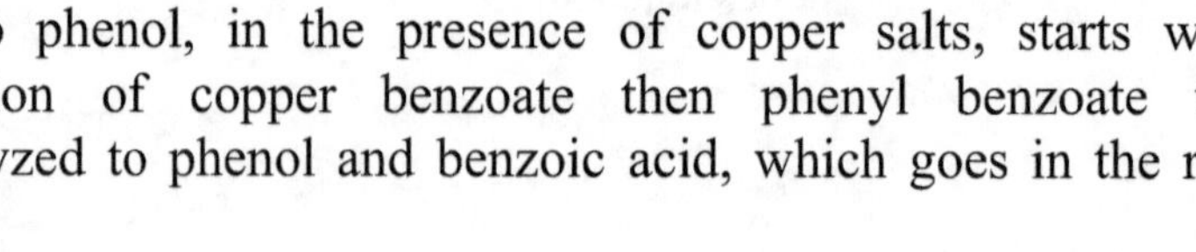

It is believed that the mechanism of reaction of oxidation of benzoic acid to phenol, in the presence of copper salts, starts with the formation of copper benzoate then phenyl benzoate that is hydrolyzed to phenol and benzoic acid, which goes in the reaction again:

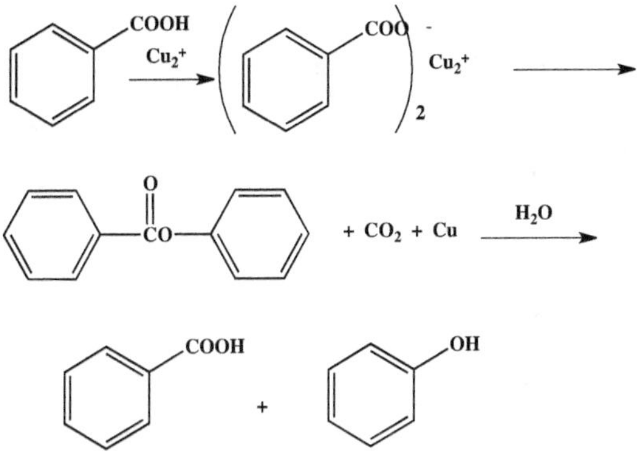

Phenol is synthesized conventionally from benzene by some other outdated procedures that are still under use in some poor countries in the developing World, for example, sulfonation that becomes meaningful especially in small enterprises with demand for the sodium sulfite and sodium bisulfite byproducts obtained:

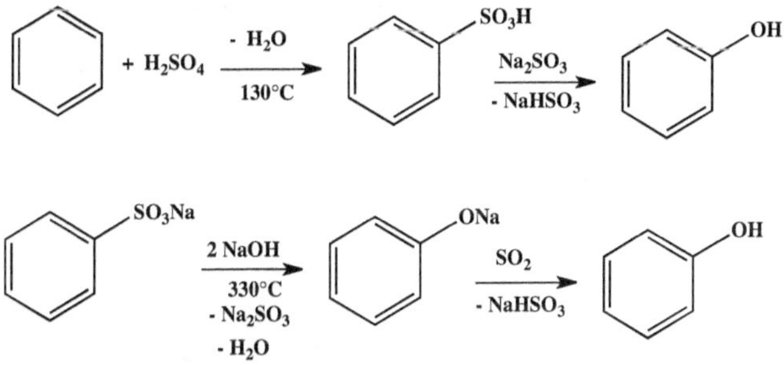

Phenol can be prepared through chlorination by converting benzene to chlorobenzene first, and then treatment with caustic soda under high temperature and pressure.

Alternatively, phenol can be prepared by reaction of benzene with water directly at 500°C, in the presence of a calcium phosphate catalyst over silica, as shown in the following equations. This method is satisfactory if the possibility of converting the sodium chloride produced into chlorine and caustic soda exists:

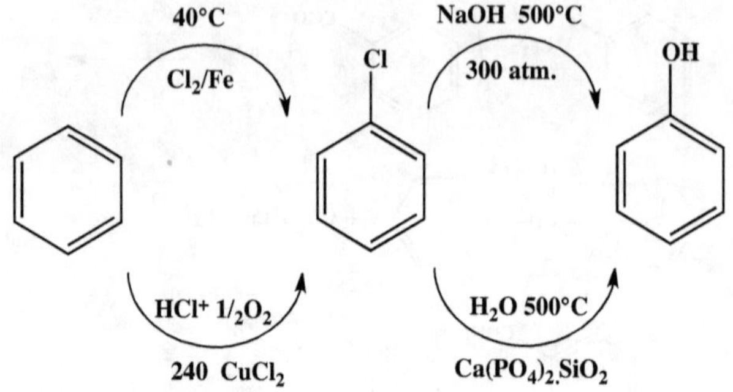

Phenol has many industrial uses worldwide; they were summarized in Figure 6-5.

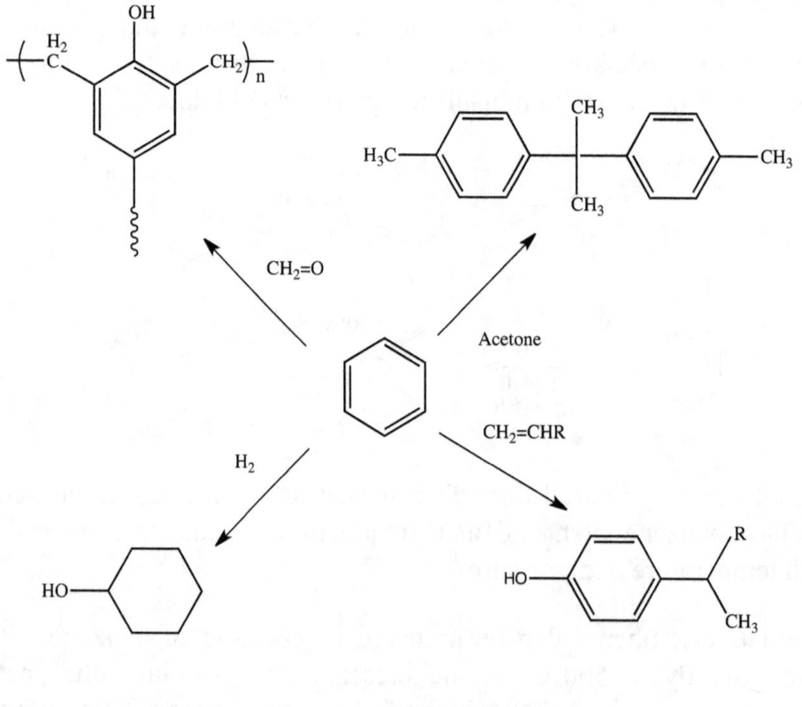

Figure 6-5: The main derivatives of phenol.

168

Nitrobenzene

Nitration is one of the oldest methods used in the manufacture of organic chemicals. It has been in use commercially for more than a century. The German scientist Mitscherlich discovered this process in 1838.

Nitration of benzene is afforded by the addition of a mixture of sulfuric acid (62%), nitric acid (30%) and water (8%) to benzene at 50°C over two to four hours. The yield is 95-98% of nitrobenzene:

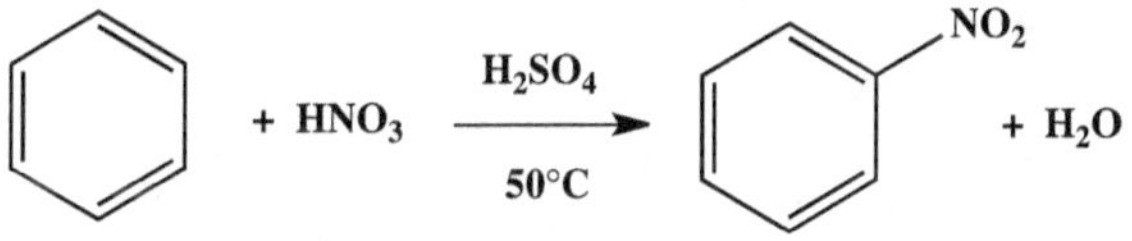

Hydrogenation of nitrobenzene is the best principal method of preparation of aniline. Aniline is used in the manufacture of dyes. Nitrobenzene is reduced in the gas phase at 270°C, and under a pressure slightly higher than atmospheric pressure, and an excess of hydrogen, in the presence of a catalyst of copper on silica. The yield is higher than 98%. Nitrobenzene can be hydrogenated in the liquid phase in the presence of Raney nickel catalyst, as well as in the presence of hydrochloric acid and iron. The iron oxides obtained as a byproduct are used in the pigments industry.

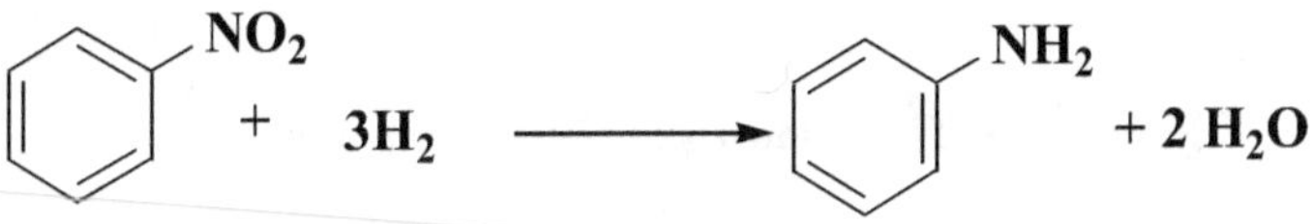

Aniline is produced in small quantities through the reaction of chlorobenzene with ammonia gas, in the presence of cuprous salts catalyst, at 210°C, and under a pressure of 60-70 atmospheres, and an excess of ammonia. The yield is 90%:

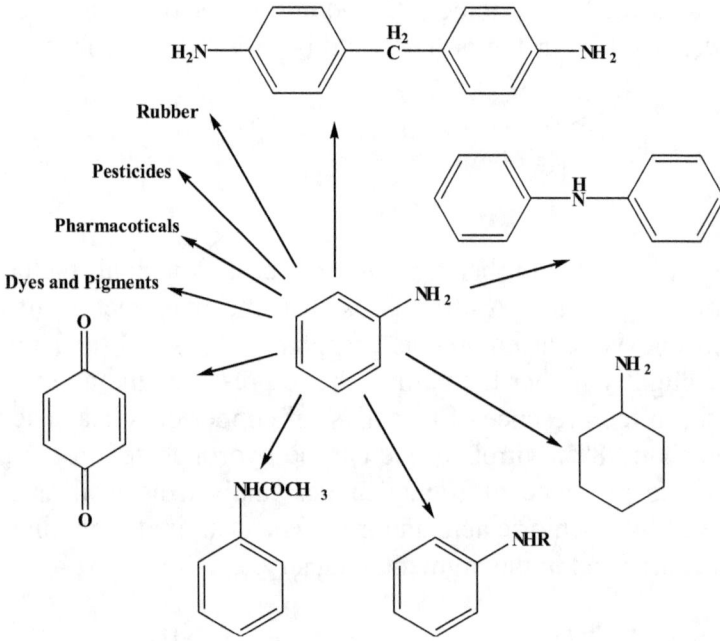

Figure (6-6) represents the most essential uses of aniline.

Figure 6-6: The utmost significant uses of aniline.

Maleic anhydride

The reaction of benzene with air for the syntheses of maleic anhydride is exothermic and the heat produced is used for manufacturing high-pressure steam. Vanadium pentoxide is used as a catalyst at 500°C:

170

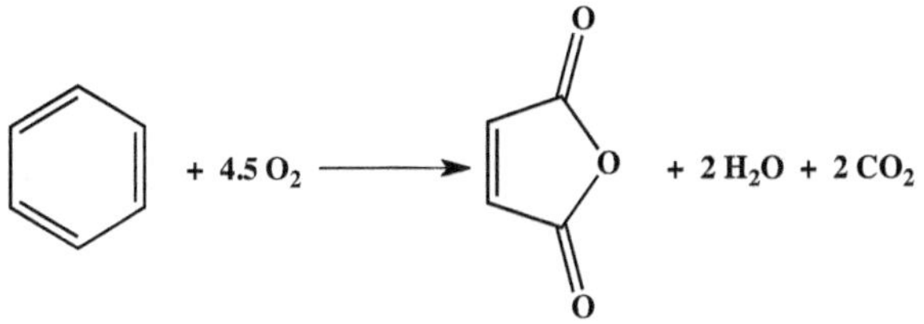

It is noted in this reaction that a large amount of carbon dioxide (30%) is produced together with maleic anhydride (70%). Carbon dioxide is used in the refrigeration and preservation industry.

Maleic anhydride finds numerous uses in the manufacture of resins, formic acid, maleic acid, and tartaric acid. The synthesis of maleic anhydride from butenes has been discussed previously.

n-Butane attracted the attention of the global manufacturers and its demand is at increase for maleic anhydride production, due to the excessive price of benzene, and the increased demand for n-butene for other purposes. n-Butane was used to obtain maleic anhydride, at similar conditions, however the yield does not exceed 10-15%.

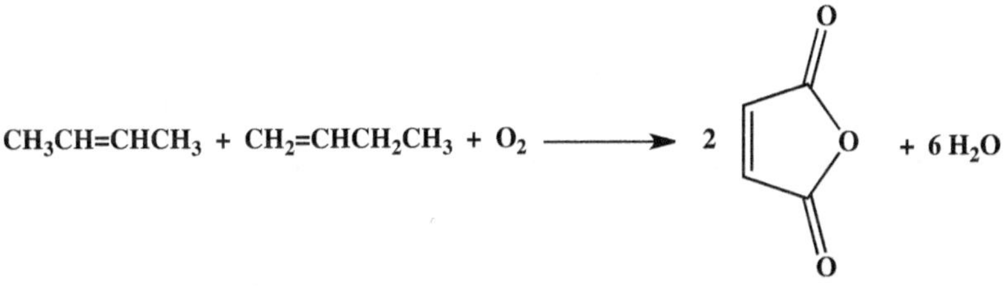

Figure (6-7) shows the main principal products derived from maleic anhydride.

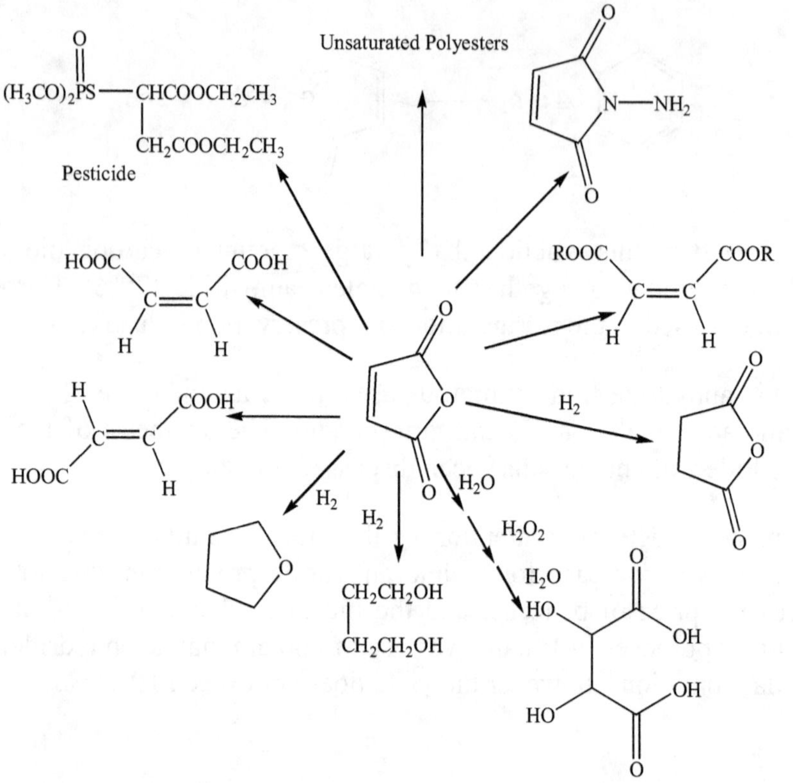

Figure 6-7: The principal products derived from maleic anhydride.

PETROCHEMICALS FROM TOLUENE

Figure 6-8 illustrates the greatest significant industrial derivatives of toluene. Table (6-9) summarizes the global distribution of uses of toluene in 2010.

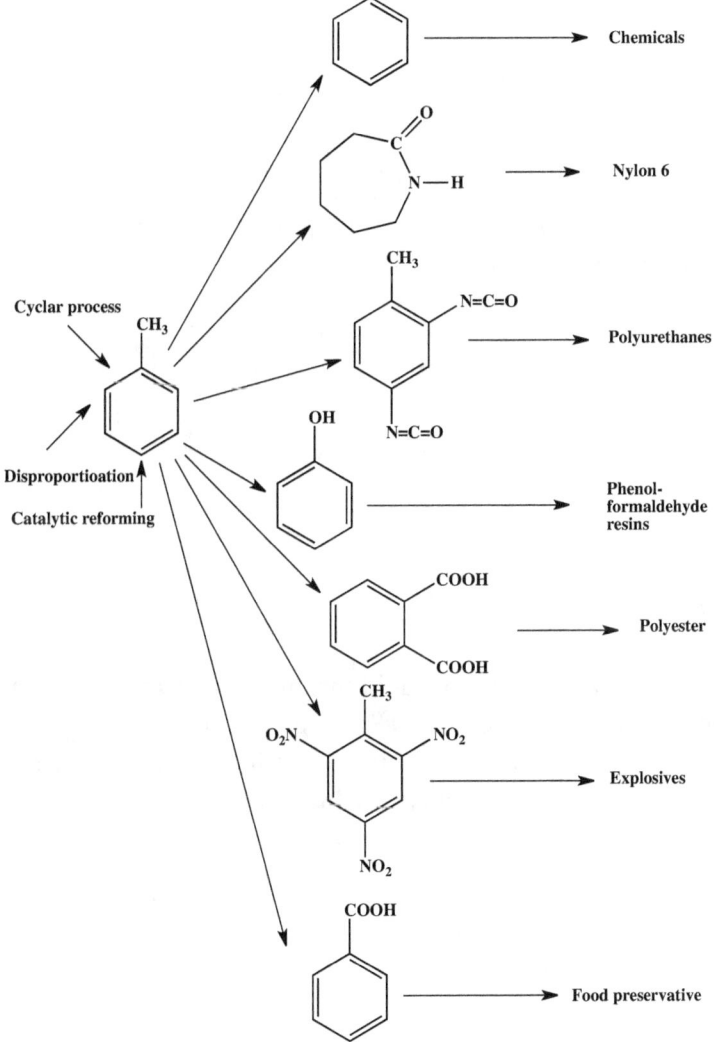

Figure 6-8: The utmost essential industrial derivatives of toluene.

Table 6-9: The typical global distribution of toluene uses, 2010.

Type of use	%
Gasoline blending	5.0
Toluene diisocyanate	5.8
Solvent	20.0
Xylenes	27.6
Benzene	29.7
Other	11.9

The details of the synthesis of toluene diisocyanate from toluene have been mentioned elsewhere in the previous chapters. A summary is outlined in the following equation:

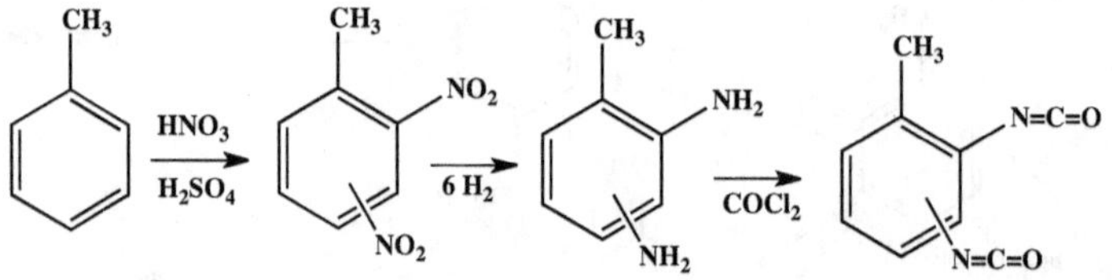

Benzoic Acid

Toluene is oxidized to benzoic acid through various methods. Air, for example, oxidizes toluene at 150°C, in the presence of cobalt salts as a catalyst to give benzoic acid in 90% yield. This method is preferable compared to the other methods. Benzoic acid is used in the manufacture of various benzoates as food preservatives.

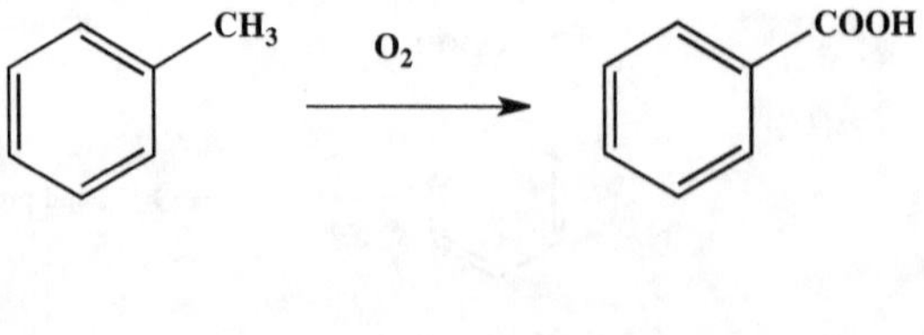

174

Trinitrotoluene

The nitration of toluene gives mononitrotoluene, dinitrotoluene, and trinitrotoluene. Trinitrotoluene (TNT) is used in the military explosives. Dinitrotoluene is used in the manufacture of toluene diisocyanate; the famous intermediate in the manufacture of polyurethane foams. Mononitrotoluene is used in the manufacture of toluidine and some other dyestuffs intermediates. As a result of the activation role played by the methyl group of toluene it can be nitrated much faster and easier than benzene.

For the preparation of trinitrotoluene, toluene is nitrated in three steps and the last step requires a larger concentration of nitric acid and sulfuric acid because of the existence of the two deactivating nitro groups on the benzene ring:

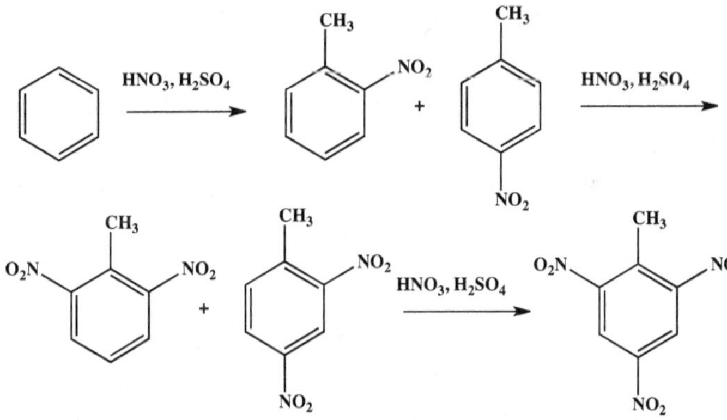

ε-caprolactam

The synthesis of ε-caprolactam starts from air oxidation of pure toluene at 160°C to benzoic acid, in the presence of a cobalt salt catalyst and a pressure of ten atmospheres as mentioned before. Then benzoic acid is hydrogenated to hexahydrobenzioc acid at 150°C, and under the pressure of ten atmospheres, in the presence of a palladium catalyst (5%) on carbon:

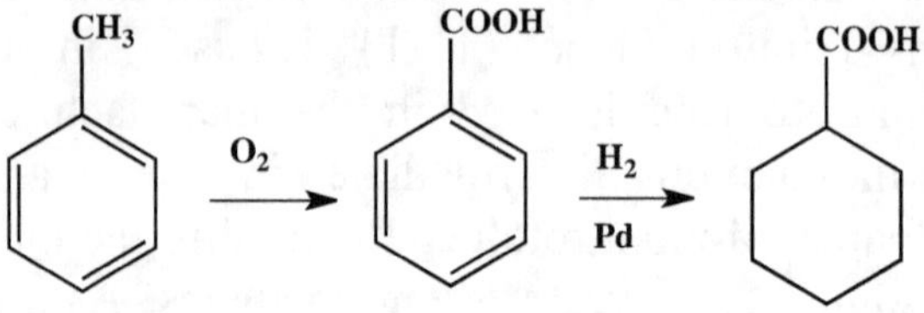

The reaction of hexahydrobenzioc acid with nitrosylsulfuric acid in oleum ($H_2SO_4.SO_3$) at 80°C gives ε-caprolactam with the loss of carbon dioxide:

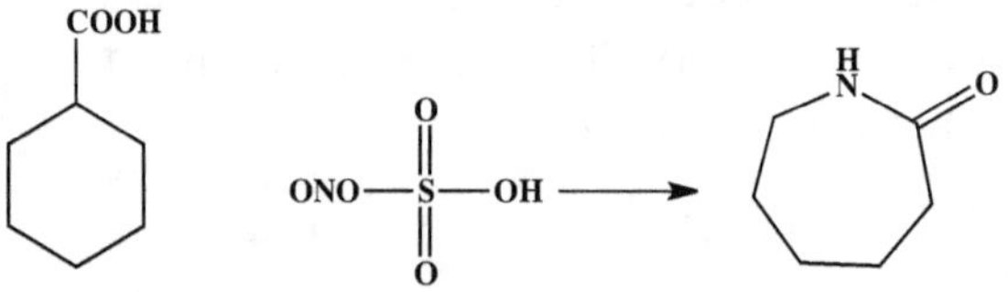

The product of this reaction contains 30% benzoic acid in toluene with a few other by-products. Fractional distillation is utilized to separate the un-reacted toluene from the pure benzoic acid for recycling and a heavy portion is withdrawn from the bottom of the tower as a by-product.

Benzoic acid, when hydrogenated in the presence of palladium catalyst, in a series of mobile phase reactors, at 170°C, and 16 atmospheric pressures, the conversion is complete in a single pass:

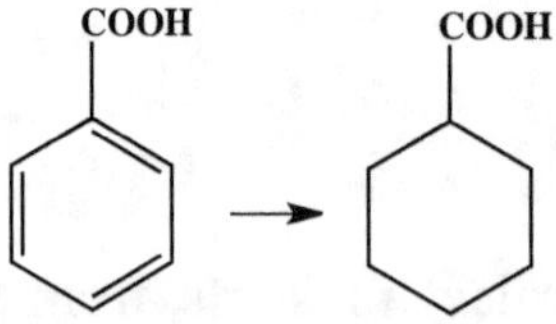

Nitrosylsulfuric acid is produced in ammonia oxidation factories where nitrogen oxides are absorbed in oleum. The reactor outputs are diluted with water. The cyclohexanecarboxylic acid is recycled to the

176

process again, while ε-caprolactam flows to the crystallization section, where it is neutralized with ammonia. Pure white ammonium sulfate crystallizes, and taken as a useful by-product at the bottom. ε-caprolactam is withdrawn from the top, for purification by solvent-solvent extraction (toluene and water) and continuous fractional distillation.

Many plants are using this SNIA, BPD, SPA, technology were established with a total production capacity of hundreds of tons per year.

ε-caprolactam is also produced through the hydrogenation of benzene to cyclohexane, followed by air oxidation to the cyclohexanone. The latter is then reacted with hydroxylamine to afford the oxime that is then treated with sulfuric acid to rearrange to the ε-caprolactam. This is reacted by ring opening polymerization to Nylon 6 as illustrated in the following scheme:

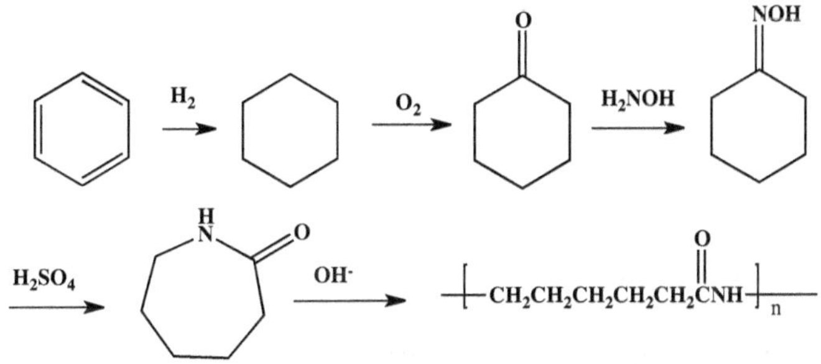

PETROCHEMICALS FROM XYLENES

Ortho- and p-xylenes have numerous appropriate industrial applications, such as phthalic anhydride from o-xylene and terephthalic acid from p-xylene. There are no appropriate industrial applications for m-xylene other than addition to motor gasoline to enhance its octane rating and its use as a solvent.

Some companies used to convert m-xylene into o- and p-xylene, at 50°C, in the presence of platinum catalyst on silica and alumina. The process is complete after the separation of ethyl benzene, o- and p-xylene from the aromatic C_8 fraction. In this way, the expensive o- and p-xylenes are produced from cheap m-xylene:

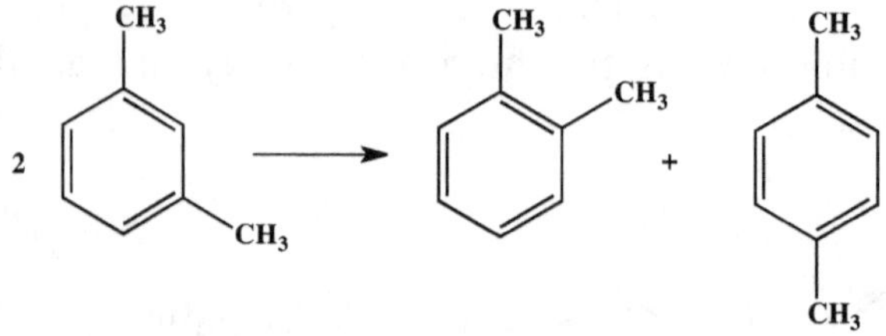

Isomerization of Xylenes

The C_8 oil fraction contains ethyl benzene in addition to ortho, meta and p-xylene. m-Xylene barely shows any visible industrial benefit and the easiest way to get rid of it is the addition to motor gasoline to raise the octane number. However, the demand is significantly high for p-xylene as the first step in the manufacture of the famous polyester (polyethylene terephthalate). o-Xylene is used intensively as well in the manufacture of phthalic anhydride. The latter is essential in the manufacture of some types of resins and paint materials, in addition to its utmost importance in the manufacture of the eminent plasticizer: dioctyl phthalate that is used frequently and in abundance in various plastics industries.

178

UOP owns an excellent technology, known as Isomar, to convert ethyl benzene or xylenes to benzene with high efficiency. In another process called Parex, UOP produced a reactor for the production and separation of p-xylene in 99.99% purity from C_8 oil fraction mixture, and the separation efficiency exceeds 97% each time. Table (6-10) illustrates the yields and feeds for a complex containing the above processes.

Table 6-10: The typical distribution of the yields of Isomar and Parex complex.

Composition	Products wt%	Fresh feed wt%
Ethylbenzene		25.5
p-Xylene	71.1	14
m-Xylene		41
o-Xylene	19.6	19.5

Phthalic Anhydride

The entire world used to get its needs of phthalic anhydride from the oxidation of naphthalene, which is obtained from the destructive distillation of coal before 1960. The same is happening in the preparation of maleic anhydride from benzene, and almost under the same conditions. Phthalic anhydride can be obtained through the oxidation of naphthalene, at 400°C, over a catalyst of vanadium pentoxide, but naphthalene is scarce nowadays, and the reaction is very exothermic, and also two carbons are lost by oxidation to carbon dioxide:

179

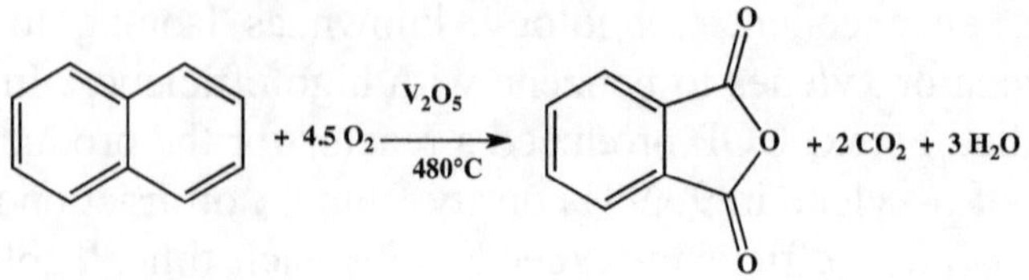

However, with the beginning of the twenty-first century, the World is getting essentially more than 90% of its needs of phthalic anhydride from o-xylene.

o-Xylene is oxidized by air to phthalic anhydride over a catalyst of vanadium pentoxide, at 550°C in the gas phase:

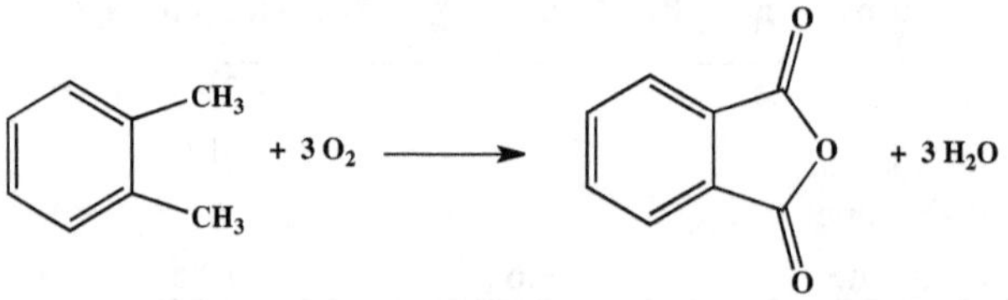

The following compounds represent the by-products of oxidation of o-xylene. All the byproducts are characterized by significant and wide industrial uses:

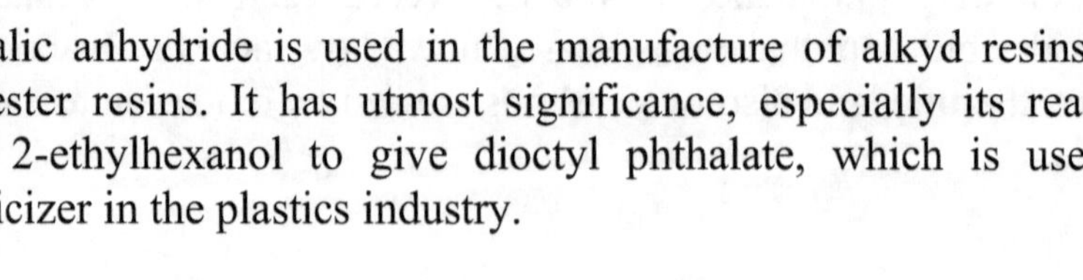

Phthalic anhydride is used in the manufacture of alkyd resins and polyester resins. It has utmost significance, especially its reaction with 2-ethylhexanol to give dioctyl phthalate, which is used as plasticizer in the plastics industry.

180

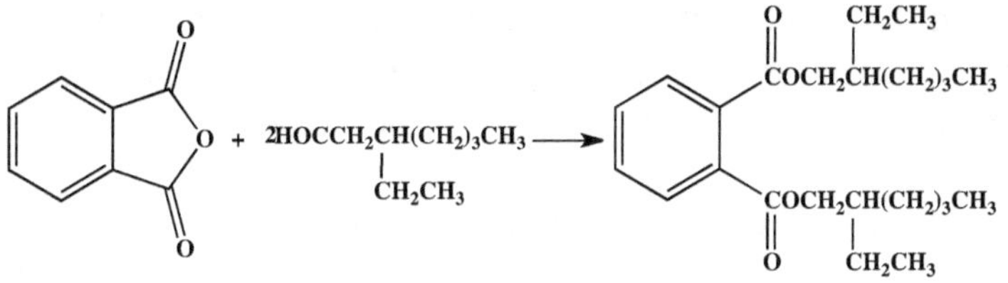

Isophthalic acid is synthesized by the oxidation of m-xylene, which is in high demand for the manufacture of polybenzamidazole known for its high thermal stability as a plastic needed in various industries such as the manufacture of small plastic airplanes.

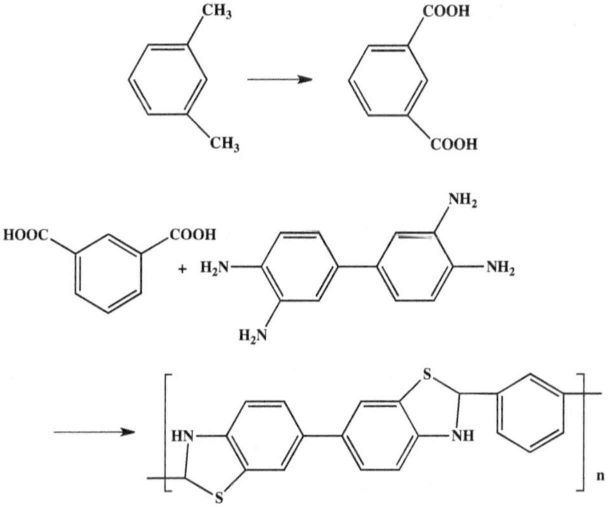

Terephthalic Acid

Pure p-xylene is oxidized to terephthalic acid, which is used as a basic starting material in the manufacture of polyethylene terephthalate, known commercially as Dacron and used in the manufacture of fabrics.

p-Xylene is oxidized with nitric acid at 150-200°C under atmospheric pressure. Although this method is still in use, however it suffers from the consumption of a large quantity of nitric acid, which is relatively expensive, besides that it produces some undesirable byproducts that are difficult to separate and dispose.

p-Xylene is oxidized with oxygen in a single step at 130°C, and under a pressure of 15 atmospheres, in the presence of acetic acid as a solvent. A cobalt salts catalyst that is soluble in acetic acid, such as cobalt acetate, is necessary. Ethyl methyl ketone is used as a catalyst activator:

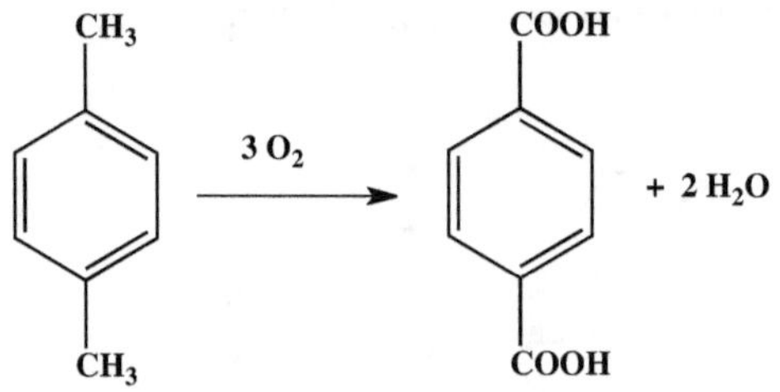

It should be noted that the required acid or ester must have a very high degree of purity, approaching 100%. The purification of the ester is much easier than the acid.

This reaction proceeds in the liquid phase in the presence of acetic acid as a solvent. The resulting terephthalic acid is in the purity requested by the desirable end products. Fiber weaving and spinning industry requires terephthalic acid with very high purity, approaching 100%, to be able reach the high molecular weights required for the properties and specifications suitable for the production of industrial fibers and threads. The terephthalic acid produced with less purity is suitable for the production of engineering plastics and packaging materials such as plastic bottles and plastic food containers, films, and some other different fibers.

Lurgi Oil-Gas Chemie GmbH specialized in the production of these technologies in collaboration with Eastman Company.

Many companies considered the cheap toluene as a starting material for the manufacture of terephthalic acid and some good opinions

182

were developed in that direction. When toluene was oxidized with air at 120°C, and under a pressure of 3 atmospheres, in the presence of a cobalt catalyst, benzoic acid was obtained. Potassium benzoate, at 400°C and under a pressure of carbon dioxide in the presence of cadmium and zinc salts as catalyst, gave potassium terephthalate, which when acidified with sulfuric acid afforded terephthalic acid and potassium sulfate that can be marketed as a fertilizer:

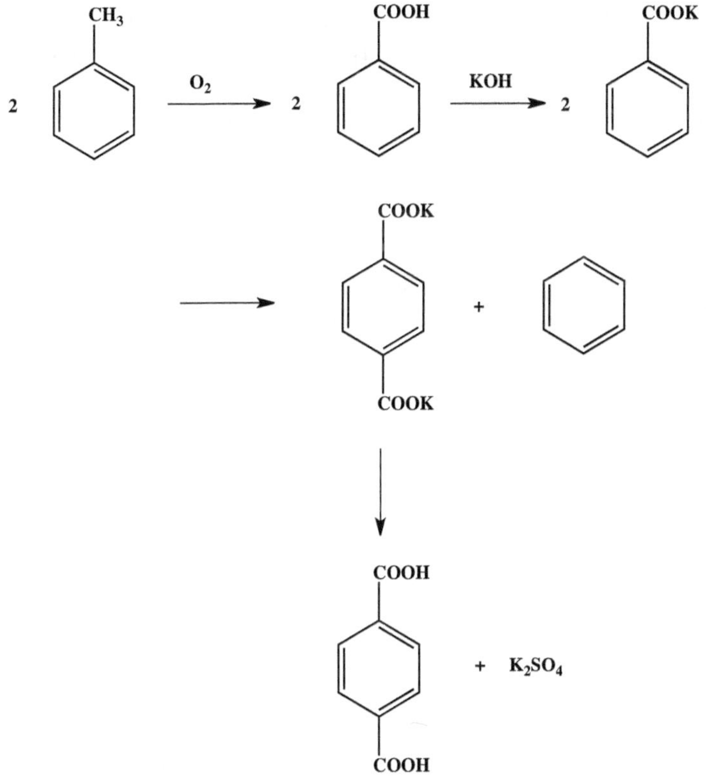

Phthalic acid can be used for the same purpose:

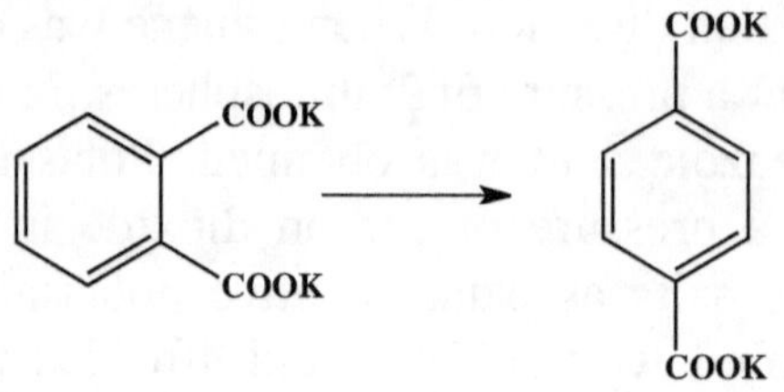

Terephthalic acid can also be synthesized from the hydrolysis of dimethyl terephthalate as seen in the following equation, which in turn is made from terephthalic acid and methanol, however the aster can be purified before hydrolysis:

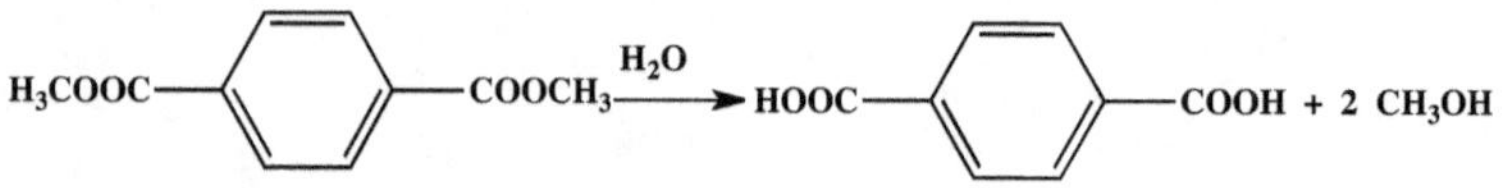

$$H_3COOC-\langle\text{benzene}\rangle-COOCH_3 \xrightarrow{H_2O} HOOC-\langle\text{benzene}\rangle-COOH + 2\ CH_3OH$$

It is well known that the production of terephthalic acid depending on p-xylene through terephthalic nitriles proceeds in two steps according to the following equations:

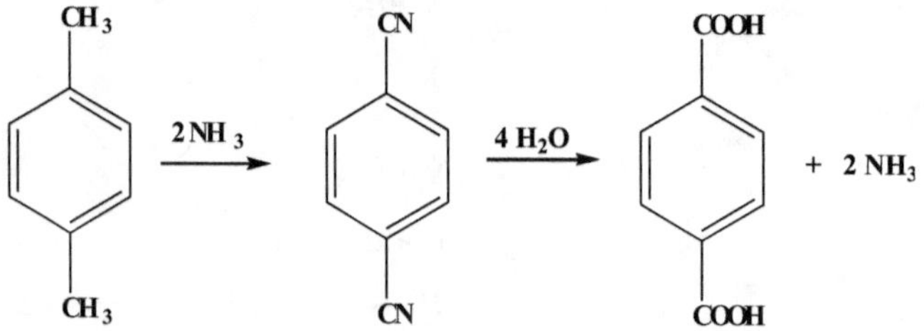

The ammoxidation of p-xylene in the gas phase in two steps characterizes this process:

1) The reaction of p-xylene with ammonia through the reduction of the oxidation catalyst (based on metal oxide).

2) The recovery of the catalyst is accomplished by oxidation in a separate step in the process. The company that owns this technology is ABB Lummus.

184

Depending on the same technology, with the desire to use the available and cheap m-xylene, Showa Denko Company tried the ammoxidation of m-xylene followed by hydrogenation for the manufacture of the diamine. The process is also applied in the case of p-xylene:

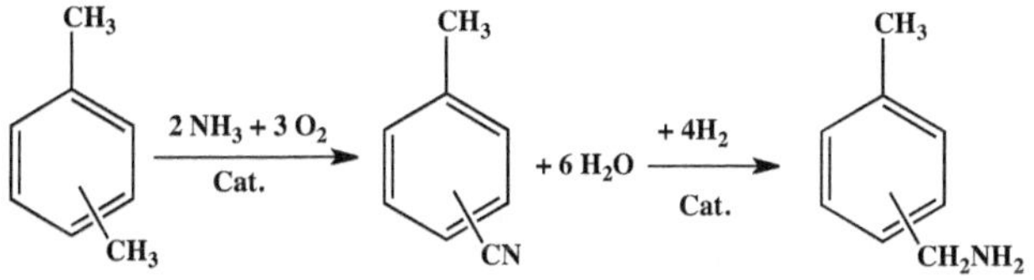

The products are ready to acquire phthalic diisocyanates, which are used in the manufacture of polyurethanes.

Mitsui Company has developed a method to manufacture terephthalic acid from the cheap and abundant toluene by its reaction with carbon monoxide at 30-40°C, in the presence of HF/BF$_3$ as a catalyst to yield p-tolualdehyde, in 96% yield, followed by purification and oxidation to terephthalic acid:

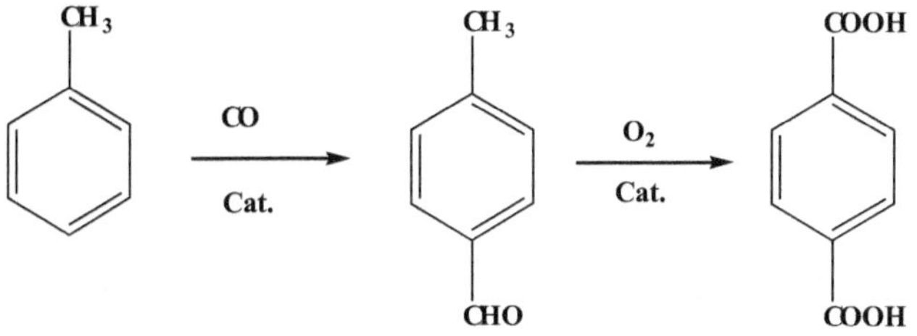

A summary of the basic consumptions of terephthalic acid is outlined as follows:

1) The esters of terephthalic acid with alcohols of 4-10 carbons.

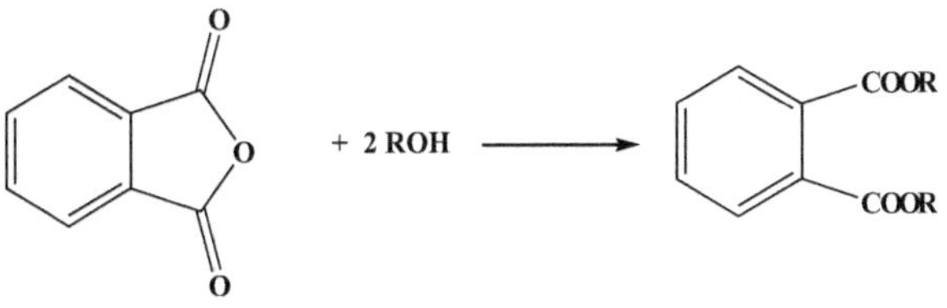

These esters are used as plasticizers in the plastics industry and the best known of them is dioctyl terephthalate.

2) The unsaturated esters are used in the manufacture of paints, elastomers, fishing fiber glass-reinforced boats and communication equipment:

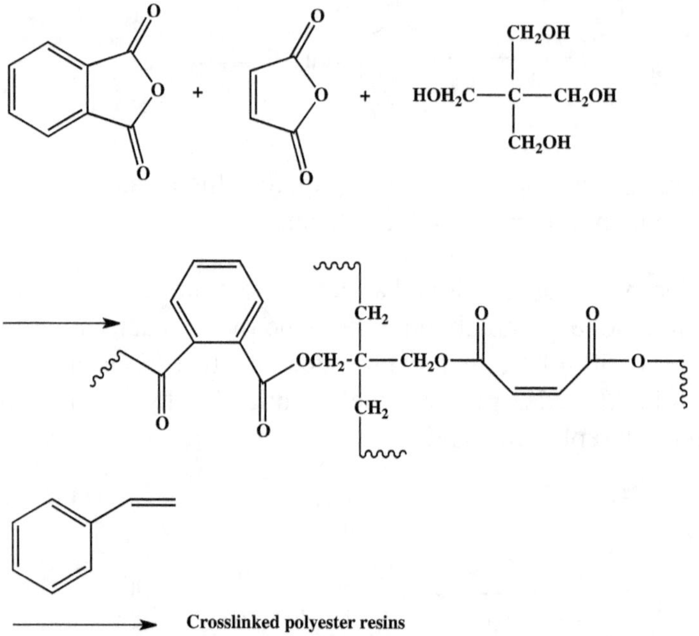

Crosslinked polyester resins

3) Polyesters of terephthalic acid with glycerin yield alkyd resins.

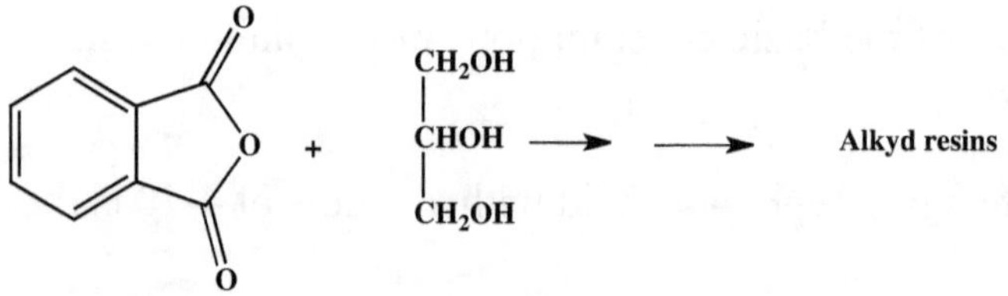

Alkyd resins

4) Other intermediate chemicals such as those used in the manufacture of dyes. Etc.

Dimethyl Terephthalate

Terephthalic acid is used in the manufacture of polyethylene terephthalate, which usually takes place in two steps. An excess of the ethylene glycol is reacted with terephthalic acid, or dimethyl terephthalate, in the first step and the ester is preferred due to the ease of purification, usually at 100-150°C, and under a pressure of 10-70 atmospheres, in the presence of a catalyst of zinc acetate and antimony trioxide, to catalyze the trans-esterification and polymerization. This is the easiest method of manufacture of the polyester, but the difficulty comes from the impossibility of acid purification, since it does not melt, nor dissolve, and sublimes at 300°C, and therefore the acid is purified in the form of the dimethyl ester:

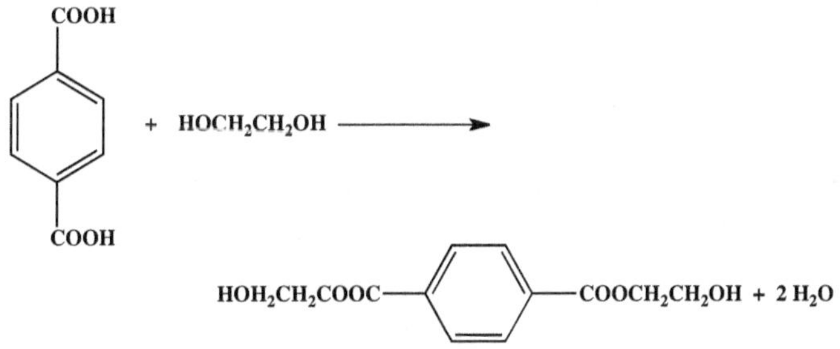

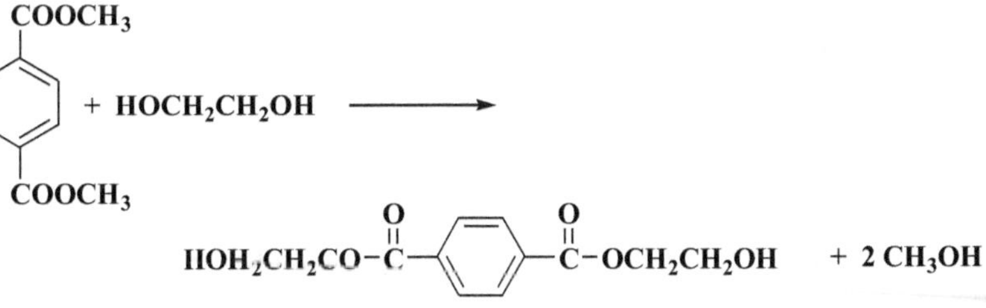

In the second step, which takes place at 260°C, an ethylene glycol molecule is expelled, and the trans-esterification continues to produce the polyethylene terephthalate. The continuous removal of ethylene glycol will help the reaction in proceedings forward. This is accomplished by passing an inert gas through the reaction, and by performing the reaction under vacuum:

187

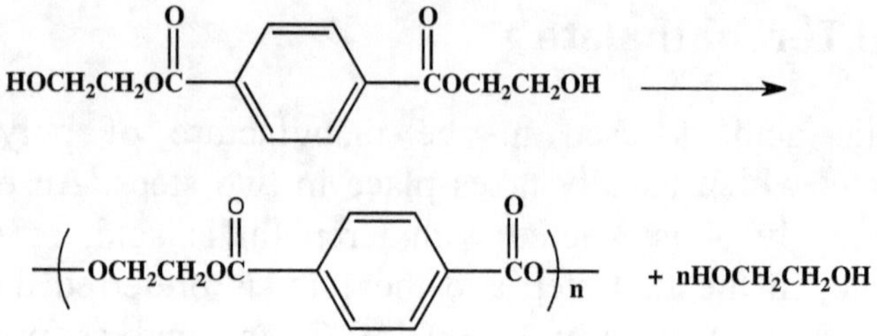

$$HOCH_2CH_2OC\overset{\displaystyle O}{\|}-\bigcirc-C\overset{\displaystyle O}{\|}OCH_2CH_2OH \longrightarrow$$

$$\left(OCH_2CH_2OC\overset{\displaystyle O}{\|}-\bigcirc-C\overset{\displaystyle O}{\|}O\right)_n + nHOCH_2CH_2OH$$

Methyl terephthalate is synthesized from p-xylene and methanol. p-Xylene is oxidized with air in the presence of a heavy metal catalyst at 140-170°C and a pressure of 4-8 atmospheres:

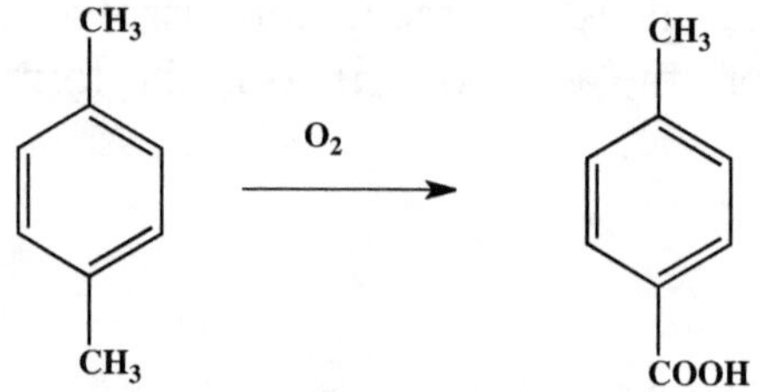

In the presence of methanol, p-tolueic acid is converted into the ester at 250-280°C, and a pressure of 20-25 atmospheres:

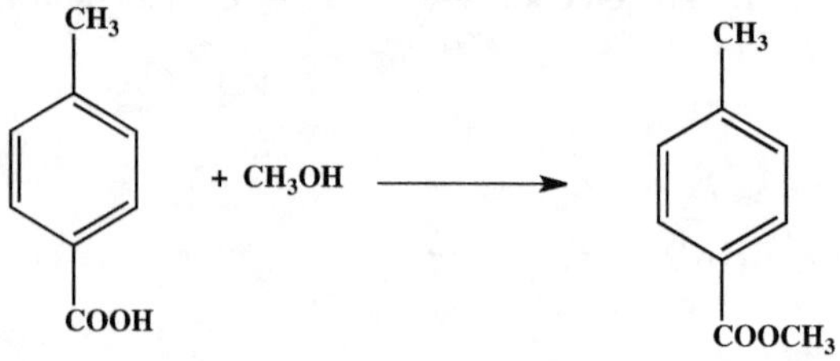

At the same previous conditions, p-methyl toluate is oxidized to the acid:

188

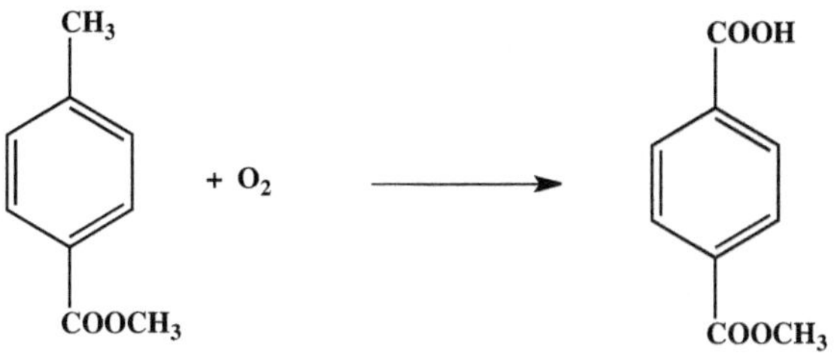

CH3 — [benzene ring] — COOCH3   + O2   ⟶   COOH — [benzene ring] — COOCH3

Then the product is reacted with methanol to produce dimethyl terephthalate:

COOH — [benzene ring] — COOCH3   + CH3OH   ⟶   COOCH3 — [benzene ring] — COOCH3

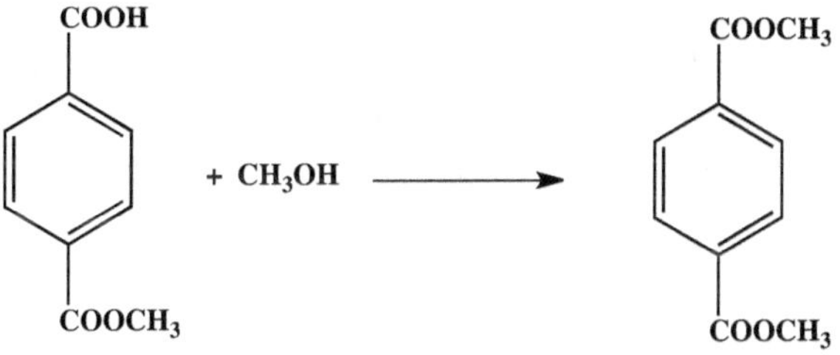

This reaction does not require the presence of a solvent through oxidation, and this spares the subsequent separation operations. Oxidation of p-xylene to p-methyltoluate is a continuous process.

The oxidation products are esterified by methanol. The excess methanol is distilled and recycled. The raw ester is withdrawn from the reactor bottom, and distilled in a column where p-methyl toluate is separated, to be recycled for complete oxidation of what emerges from the bottom of the column. This is separated in another column to raw dimethyl terephthalate. The remnants are withdrawn from the bottom.

Dimethyl terephthalate is subjected to further purification by crystallization from methanol. The product is separated from methanol by centrifuge, and the filtrate is distilled in a column, where methanol separates from the top for recycling to the methanol tank. The isomers were removed from the bottom.

The molten dimethyl terephthalate could be pumped directly to the polymerization process or transferred by tank cars to the factories of glass-reinforced plastics.

De Gussa owns this technology, which has been sold to tens of factories around the World, producing many million tons per year of pure dimethyl terephthalate.

To avoid trans-esterification of dimethyl terephthalate with ethylene glycol, due to the difficulty of the direct reaction of terephthalic acid with the ethylene glycol, attempts were made to react the acid with ethylene oxide directly in solution at 90-130°C, and under a pressure of 20-30 atmospheres, in the presence of a catalyst of amine or tertiary alkyl aluminum salts, to obtain dihydroxy ethyl terephthalate:

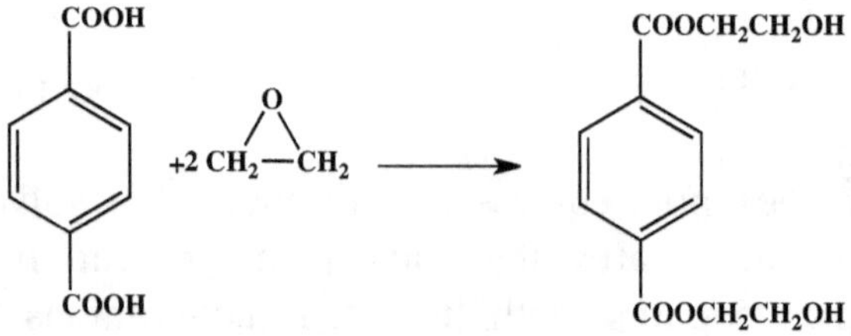

There are enormous advantages for this intermediate, including the possibility of recrystallization from water, where terephthalic acid needs no purification prior to this step, in addition to the advantage of the absence of water or methanol to get rid of, and furthermore the reaction is faster.

Among the interesting uses of dimethyl terephthalate is its hydrogenation in two steps; the first at 160-180°C, and under 300-400 atmospheres of pressure, in the presence of palladium catalyst, to give cis and trans cyclohexane dimethyl carboxylic ester, in 97% yield. In the second step the ester is hydrogenated in the presence of a copper chromate catalyst, without the need for previous purification:

This product is used as a dialcohol in the manufacture of polyesters, polyurethanes and polycarbonates.

References (General)

1. K. Weissermel, H.-J. Arpe, Industrial Organic Chemistry, Wiley-VCH, 4th Ed., 2003.

2. Shell, Chemicals Information Handbook, 1986.

A. M. Brownstein, Trends in Petrochemical Technology, Petroleum Publishing Company, Tulsa, Oklahoma, 1976.

3. Shell, The Petroleum Handbook, Elsevier, 1983.

4. British Petroleum, Our Industry Petroleum, 1977.

5. Peter Wiseman, Industrial Organic Chemistry, 2nd Ed., Applied Science, 1983.

6. Peter Wiseman, Petrochemicals, UMIST, 1986.

A. Lawrence Waddams, Chemicals from Petroleum, 4th Ed. Murray, London, 1978.

7. G. Austin, Shreve's Chemical Process Industries, 5th Ed., McGraw-Hill, 1984.

8. G. D. Hobson and W. Pohl, Modern Petroleum Technology, 4th Ed. Applied Science, London, 1973.

9. K. J. Saunders, Organic Polymer Chemistry, Chapman and Hall, London, 1973.

10. C. A. Clausen III and G. Mattson, Principles of Industrial Chemistry, Wiley, 1978.

11. M. P. Stevens, Polymer Chemistry, Addison-Wesley, 1975.

12. F. W. Billmeyer, Jr., Textbook of Polymer Science, Wiley, 1984.

13. N. N. Lebedv, Chemistry and Technology of Basic Organic and Petrochemical Synthesis, Vol. 1 &2, Mir, 1984.

14. OPEC Recent Publications.

15. OAPEC Recent Publications.

16. Chemicals and Engineering News, Recent Issues.

17. Hydrocarbon Processing and Petroleum Refiner, Recent Issues.

18. European Chemical News, Recent Issues.

19. Chemical Week, Recent Issues.

20. Chemical Market Reporter, Recent Issues.

www.ingramcontent.com/pod-product-compliance
Lightning Source LLC
Chambersburg PA
CBHW051457170526
45166CB00001B/288